Jack Pearce
Fundamentals for the Anthropocene

Jack Pearce

Fundamentals for the Anthropocene

Managing Editor: Paulina Leśna-Szreter

DE GRUYTER

ISBN 978-3-11-056730-4
e- ISBN 978-3-11-056730-4
ISBN EPUB 978-3-11-056730-4

Library of Congress Cataloging-in-Publication Data
A CIP catalog record for this book has been applied for at the Library of Congress.

© 2017 Jack Pearce
Published by De Gruyter Open Ltd, Warsaw/Berlin
Part of Walter de Gruyter GmbH, Berlin/Boston
The book is published with open access at www.degruyter.com.

Managing Editor: Paulina Leśna-Szreter

www.degruyteropen.com
Cover illustration: © Ingram Publishing, Thinkstock

Contents

Introduction

At dinner at Star Island, during a 2008 conference on "Emergence" convened by the Institute on Religion in an Age of Science, Paul Cassell exclaimed, somewhat incredulously, as I recall, "Are you saying that all this (the Universe) is just the result of energy minimization?"[1]

Pretty close. More precisely, I tried to suggest, as well as I could at dinner table, that one can see that the visible, tangible Universe is the set of continuing, progressive correlations between interacting elements, forming systems of relationships. This occurs in a continuing cascade of phase transitions universe-wide. This cascade of phase transitions toward greater correlation of interacting elements, overall, in the universe as a whole, creates both an "arrow of time" and what we call visible "order" in the Universe. And, back to Paul's question, it involves energy minimization in forming the relationships -- the correlations.

This is a rather abstract theme, I grant. But it ties together, runs through, and has powerful explanatory value as to many different phenomena, and many fields of study – including hierarchy theory, systems theory, power law distributions, social networks, "emergence" theory, structure in biological systems, and more.

Surprisingly, for something this abstract, this line of thinking, developed in a step by step fashion, can also give us insight into challenges which face us daily – how we create and maintain the organizations which sustain us and give us our human potentials, what languages are and how they work, the nature of ethics, and how we might try to conceive our human objectives, possibilities, and social necessities.

Another necessary prelude to the architecture of concepts here set out is that what we call the universe is a framework sustained by and expressive of what we characterize as energy flows.

The substance of this realization goes back at least as far as those Greek philosophers who saw all things as process. Currently, it is often elaborated perceptively in "process philosophy". But the best comprehensive, quantitative, detailed and elaborated exploration of this understanding, in physical terms, known to me, is found in the works of the cosmologist Eric Chaisson. I will refer to him often.

Another way of expressing this understanding is to see the Universe as a set of what one can picture as "standing waves". In this way of looking at things, the whole visible (to us) Universe is a manifestation of wave-particle duality.

---

1 Though arriving at the themes set out here has been a decades-long journey, drawing on diverse sources, a couple of conferences staged by the Institute On Religion in an Age of Science, dealing with "emergence", were timely and useful in bringing together people and materials which "nurtured the concept", so to speak.

Seeing the Universe as "standing waves" allows one to use the concepts of "non-equilibrium thermodynamics".[2] That is, one can see a stable phenomenon (such as a galaxy, tree, or animal) as embodying energy flows in a stable pattern. We can also call this pattern a correlation set. This facilitates many insights.

I do not here attempt to run down, and compare and contrast, all major authors and the vast field of variants of thought which might be said to relate to issues here brought into focus. This is not done out of disrespect for the many able researchers and thinkers who have addressed the topics I address, in one fashion or another. The plan for this book is to suggest a single, unifying theme for the generation of order in the tangible universe as we encounter it, and carry it through into the realm of human affairs, in a spare, economical way. My overriding objective is to present a clean and consistent framework as of thought as simply as is possible. Trying to accompany this effort with a history of relevant thoughts would be an unwieldy enterprise.

I will, of course, cite researchers whose work has struck close to the bones of the skeleton I attempt to lay out. Their work has directly underlain and stimulated my efforts to construct this framework. Key insights of theirs can be economically explained.

In the first chapter we will start with linking the fundamental theme of correlation with basic observables in the Universe. In the second chapter we will expand on the linkage between correlation and differentiation. In the third, I introduce basic characteristics of the ordering process, and how they work together. Then I will show how these tools of thought apply to "emergence", and hierarchy theories.

Then we will deal with some often discussed issues or topics which are simplified by the framework here set up. These topics include complexity, causation, and the 'arrow of time'. These may seem to be off to the side of the main argument. But they introduce concepts which usefully clarify issues which get involved in our thinking about how life works.

Then we will take up the nature of the life process and where we fit into it.

Finally, we will move on to widely recognized challenges facing us talking apes on this Earth, both now and in the future, including the existential challenges of our future energy supply and "globalization", and what sort of "ethics" seem to be required to deal with these challenges.

---

2 The non equilibrium thermodynamics term can be a bit confusing. A stable pattern of energy flows can have -- does have -- equilibria. It has to, to exist. But in the development of physics, the concept of equilibria originated in situations which gave an appearance of repose, or stillness, or a static state. When using this way of viewing phenomena, people had to imagine how that "equilibrium" might change if movement, or energy, entered into what was seen as a static system. This led to the use of the "non-equilibrium" term, and thinking in terms of non-isolated systems.

But if all stationary, or correlated, states embody energy flows, one just deals with equilibria as equilibria, in thermodynamics. (Actually, the "thermodynamics" term itself is a bit archaic, as dynamics imply what we call energy, and energy can be expressed in more ways than what we see as temperature, or thermal energy.)

Before we begin, let me make another hopeful suggestion for the intrepid reader. I have found in exposing drafts of this manuscript to friends that the first four chapters or so have seemed to some to be unfamiliar and a bit daunting. This may be a fault of my writing style. But also it seems to me that this may at least in part be because though the language of correlation, differentiation, and hierarchies is often used in everyday life, here these themes are presented as architectural elements of a worldview. This entails a shift in perspectives on such concepts. So if you will persist through this architectural use of this terminology, you will, I hope, be rewarded in the latter chapters.

1 The Visible Universe Starts With – Believe It or Not – Correlation

The purpose of the initial chapters in this book is to depict the visible, tangible universe[3] as systems of relationships, given their substance by correlations among the elements within them. Each of such systems encounters, in external relations, a combination of randomness and correlations. Each distinguishable system is given its "meaning" by its interactions with other relational systems. All evolve against a background of randomness, over universal time, or process, to date.

This conceptual structure reaches down to the quantum mechanics level. Key features of the evolution of the Universe at what we call the "macroscopic" levels are rooted there, and manifested throughout the ordering of the Universe.

Keeping these things in mind simplifies a number of issues encountered in our attempts to understand and explain the Universe.

This "relational" view of the Universe requires us to identify how relationships arise, and are operative. This involves coming to grips with the concept of "correlation", and how correlation builds ordered states, or complexes of relationships.

Let us then focus of "correlation", and what it means in the Universe.

At the outset, we are not focusing here on the commonly used statistical means of sorting out what factors seem to be related to each other in an often complex field of "variables", or factors. I attempt to direct your attention to physical relationships -- physical connections between systems, as distinguished from random impingements of one system upon another -- for example, planets revolving around suns, the crowding of atoms within suns, the crowding of nucleons within an atom, and the like. We are talking about situations in which one element (be it an atom, a molecule, a planet, or any other physical, tangible phenomenon) has a reduction in "degrees of freedom" vis a vis another phenomenon, or system, such that one can say there is a "relationship", or non-randomness, as between them.

In doing so, let us start with our native, inbuilt perceptions, and then probe underneath them.

So let us look in two directions for the underlying nature of order in this universe as we natively experience it. Our ancestors have done so for millennia. Now we use hard-won tools of observation and equally hard-won concepts. But, instinctively, and necessarily, to position ourselves in this Universe, we still look in these two directions.

---

3 I will treat here with "fermionic" matter, which manifests the "Pauli Exclusion Principle" -- that is, the fermions involved have "fractional integral spin", and no two such entities, assuming we so characterize them, can have precisely the same characteristics at the same time and place. Such matter includes, nucleons, or hadrons, and electrons associated with them. It is, by and large, such matter, which has a "place to stand", or volume, which makes up the things humans characterize as tangible, like suns, planets, etc.

When we have done this, let us strip away the imaginings of the past and the multitude of terminologies of more recent times to see underlying themes of organization.

One direction is the vast, awesome panoply of the visible[4] universe around us – the "heavens", the "stars", and the relationships between them. The other direction is the "elements" – the bits which relate together to make up, as we see it, all tangible[5] reality, including the stars, our planets, ourselves.

Over millennia heretofore, in our cognitive infancy, we projected on the heavens patterns familiar to our everyday lives – dippers, animals, idealizations like "Gods". The catalogue of imaginings is too various to itemize, and too fanciful and too disorganized to have value.

Recently, in a rush in centuries-time, keen observers and clear thinkers have created a picture of a vast hierarchy of structure and motion.

We picture planets moving around stars; stars grouped together, moving, as it appears to us, in slow, vast swirls around swirl-centers. We call these star swirls galaxies. Galaxies – at least some 100 billion of them in the observable universe, and intimations of a great deal more -- form galaxy clusters, journeying, as it appears to us, through a void. Even more, we now discern "super clusters", each containing

---

4 I will not in this manuscript deal in any detail with "dark matter" and "dark energy". The theses of this manuscript deal with the sort of "matter" which has interactions with other matter in a way creating relationships. Dark matter interacts with the sort of matter we experience gravitationally. This does seem to give "dark matter" a nexus with the correlation process which I here attempt to expose and trace. But we cannot as yet see whether and how dark matter participates in the relational structures built on the correlations of nucleons, embodied in the layers of aggregates of those nucleons. So I do not attempt a treatment of dark matter in the combinatorial permutations built upon atoms at this point.

As to "dark energy", this is currently described as impelling the expansion of the universe at this stage of its development. Layzer (1990) and Chaisson (2001) have postulated that the expansion of the universe creates the possibility of the order which constitutes the relational, visible Universe. Thus there is a direct link between dark energy as we now imagine it and the order building of the relational, visible universe.

The connection between "dark energy" and order building is fascinating. Though discussions of dark energy seem to assume a framework of dark energy->universe expansion->order generation, one might speculate on a different form of linkage. But little can be said at this time, at least on my part, about the connection other than that now assumed.

5 As before noted, and here reiterate, in using the term "tangible reality", I refer basically to fermion manifestations -- that is, "particles", and aggregations of particles, which have the characteristic of excluding other fermionic entities i.e. things that have a "place" and exclude other states with respect to that state. In the terms of our experience, we call this "matter".

This is not to say that the aspects of the universal fabric which mediate relationships among these things -- e.g. electromagnetic radiation, "fields", and "forces", like gravity -- are not bound up in what we perceive as the universe. Obviously they are. But we are fermionic "matter" and have relationships with other fermionic matter which has its own "place" which we can not simultaneously occupy.

thousands of galaxies, stretching many hundreds of millions of light years across space. These super clusters are arranged in filament or sheet-like structures, between which there are gigantic voids of seemingly empty space.

How shall we interpret such artifact-augmented perceptions?

Let us do a thought experiment, linking this vast picture with the elements in it. Let us stretch, in a peculiar way, our imaginations to think about a "universe" which could not exist, the better to understand what the "universe" is which does exist.

Let us imagine, let us say, a universe of atoms, not clumped in stars, or planets, but in a vast, loose gas. (We could imagine all the various atomic "elements" which now exist to be in this gaseous state – though, if we paused to think about it -- those elements which are built up of many nucleons would not exist were there only a random mass of hydrogen atoms with no structure and no coordinated directionality. So let us start with hydrogen atoms.)

The first realization which occurs -- to me, at least -- is that in this random-gas universe, "we" – the observers -- would not exist.

Stop for a moment. We, or Observers like us, consist of intricately coordinated systems, made up of a number of different types of atoms. We are not randomized gas. Further, we process information – i.e. organized, non-random impingements -- from our surroundings. So the all-gas, all-random universe has no foundation for us. There is no "observer" as we think of observers.

To sharpen the point, let us suppose we, or Observers of some "organized" sort, like us in this way, do exist. Perhaps they are somehow projected into the frame, let us say from an organized universe – even though of course this makes no sense in the random "universe"

What would we, the Intruding Observers, see, or sense?

Nothing. Only perhaps some sense of heat (agitation of the elements of our system) from high levels of kinetic effects upon elements of our systems, or cold (a lowering of internal activity pattern resultant across our internal systems from loss of radiation and lack of impacts upon us resultant from low levels of impingement and perhaps low relative velocities of the elements of the gas) from the "void".[6]

There would be no differentiations, no shapes, no points of light, no swirls, nothing discernible.

Thus the Intruding Observer -- which, remember, would have no foundation for existence -- would also have no structure, as to tangible elements, to perceive.

---

6 Let us note in addition that these expressions we use here have to do with the state of the "observer", and can exist only in terms of that state or state set. Any difference in the "characteristics" of the gas would come from effects of that gas *upon a correlated, differentiated state.* Conversely, an alteration of the aggregate effects of the gas upon that correlated state set would arise from effects of that or other correlated state sets (e.g. compression or agitation) upon the elements of the gas. In effect, as differentiations arise from randomness, a set of interactions between the differentiated states and random state sets are created. This is an act of "creation", of an Universe.

What is the core point of this ultra-simple, back to basics, exercise?

One can abstract from this thought experiment one simple, central, unavoidable realization. What we, or any observer, or organized element or set of elements in this Universe, "see" (or register, or process, or depict, or mirror, or represent) is the correlations of the elements – of the atoms in whatever is outside us.

In probabilistic terms, this means that if one element of 'matter' which obeys the "Pauli exclusion principle" and thus has a "place to stand", so to speak, excluding other fermionic matter from its "state", is registered by such an correlated Observer (and by extrapolation presumed in a given structure) another will (with high probability) also be registered. Such "matter" would have to clump and cling together -- to be correlated, in the sense of if one, then another -- for us to sense them.

The suns are just big clumps of atoms. The planets as well. The galaxies are just big swirls of sun clumps (with a lot of loose gas intermixed). And so forth up the hierarchy.

So, *in architectural terms,* what we see as the "Universe" is just correlations, the co- relations, "relationship", patterns of coordination. Non-randomness, if you will.

Now let us reverse directions and go from the vast to the infinitesimal. Let us go down to the billions of components which make up our own existence–pattern. Let us go down to the "atoms" which, as far as we know, were first imagined some two millennia or so ago, but which were definitely established in the lexicon of science only within the last two centuries. Let us go to the ur-atom, the hydrogen atom, the single quark-swirl, as we now vaguely imagine it -- the proton, the hadron.

We have discovered over the last two centuries that the early-discovered 92 "elements" – or distinguishable "atoms" – are basically just hadrons, or nucleons, squished together. In common language – or in terms of correlations -- all the "elements" are just different versions of correlated groups of nucleons.

All respectable mainstream treatments of structures made up of atoms necessarily reflect this view. This assumption is so common, so basic, that the astonishing simplicity of the matter tends to drift to the edge of consciousness. This view functions in the realm of assumptions, rather than consciously occupying the center of our conceptions.

To repeat, to drag the attention to the central issue, for present purposes, all the "elements" are concatenations, aggregates, clumping's -- that is, *correlations, and systems of relationships* -- of nucleons.[7]

The mashed together nucleons are, as we now see it, formed within stars. These aggregations occur in what might be called a form of "phase transition", under great pressure and heat. Each atom-aggregate is a fearfully tight coordination of the hadron

---

7 I forego in this discussion description of the "forces" which mediate these aggregations. They are extensively characterized elsewhere.

elements, encapsulating energies which are very high when we measure them in our relatively loosely bound, larger aggregates of molecules. We have recently, however. seen these fearful energies unleashed when we learned how to split up, or disassemble, or break apart, very large atoms, such as uranium and thorium.

And thus in the simple matter of correlating nucleons into clumps -- differing in numbers and ways of interacting -- is formed the building blocks on which the vast, complex, astonishing elaboration of further differentiated relational structures occur. And those occurrences are, again, as to us the visible, tangible Universe -- the stuff we have to shove and be shoved by, because each hadron and each combination of hadrons has its own place to stand.

As we now conceive it, the relationships between "atoms", whether simple hydrogen atoms or differentiated groups of nucleons, are mediated at close range through "electron shells". There are regularities in the way the electron shells are formed, and in how they engage with other atom-aggregates. These regularities form the basis of the table of elements, used extensively in chemistry.

But before we plunge into the intricacies built into chemistry and other fields of study, let us look again at the fundamental pattern involved. The aggregation of nucleons, or hadrons, into 92+ differentiated, distinct atomic concatenations forms the basis for the vast ensemble of combinations which can be formed of these elements. Given that you have over 92 types of atoms which can be combined in different ways, you have an enormous combinatorial field available. However, we need to note that the actualization of this combinatorial field is subject to energy invariances, and to probabilities. We can use notation systems to state vast numbers of combinations of 92 "elements". But the realizations of those combinations are bounded by the physical constraints involved in creating 'relationships', and the probabilistic – I am next going to state -- hierarchical way in which they can be actualized.

And then you can get to combinations of these combinations of elements. This is where the Universe really does its stuff. *It builds groups of groups.* In this differentiated groupings of nuclear structures, and then differentiated groupings of groupings, is realized the richness and complexity of the Universe.

For most of this book, we will proceed from this point upwards, so to speak. But we can go deeper even than the atom, into the realm of quantum mechanics. There is a reward for doing so.

There, at the "quantum mechanics" level, instead of describing "particles", and using ideas of positions and momenta of identified objects relative to each other, physicists have conceptualized a "wave function" which can be spread over a region, and which can have states "superposed".

Here our physicists speak in terms of a "density matrix" used to calculate the probabilities of occurrences of defined states, and "measurements" thereof. In this realm, our human physical intuitions as to shape and position seem inapplicable.

This has seemed to most of us to be alien territory. (Indeed, even the term "territory" may get fuzzy.) But theoreticians keep probing it.

In the last decades of the 20th and first decade of the 21s centuries, theoreticians have created a body of ideas called "relational quantum mechanics". This field was stimulated in large measure by a seminal article by Carlo Rovelli (1996).

This body of thinking puts in central focus the following: (a) all systems are "quantum systems", (b) quantum systems *"measure" each other by physical interaction,* and (c) the physical interactions, which result in "entanglement", occur by means of reduction in degrees of freedom, and hence correlation.

The term "relational" in the description of the field appears to arise from the perception that the relation of quantum systems to each other is at the core of the operation of such systems. Focusing on the relations between quantum systems – at bottom the correlations between them – helps us decipher what has been so puzzling to us about them.

That is among the contributions of Rovelli and his colleagues. The step which I take, and suggest to you, is to recognize and conceptually to build upon the proposition that relational structures are built by correlational processes here at the bottom of reality (or as near it as humans seem to have gotten) -- at what is commonly called the "quantum mechanics" level. It is hard for us to see into that level, using our accustomed frames of reference, precisely because we are made of correlations, and all we can "see" at the quantum level is the fact of correlation arising there.

From this odd, hard to grasp level of reality, those correlational processes are, as I have earlier indicated, in play upward throughout the hierarchies visible in the Universe. *If we follow the logic of Rovelli et al, we need to keep in. mind that each system, at all levels, is a "quantum" system, and "measures", and gets its "meaning"[8] from, the relationship, or interactions, with other systems which may be encountered.*

Rovelli and others also specify that the quantum level interactions can be stated in terms of information. They thus link the quantum mechanics foundation of the universe to the concepts of a computational universe. Indeed, Rovelli asserts that "Quantum mechanics is a theory about information" (p.3), and "Correlation is information in the sense of information theory..." referring to Shannon's work (p.9).

Rovelli points up connections between the Hilbert space mode of analysis, the Boltzmann-Gibbs-Shannon entropy measures, and the penetrating form of information analysis set out in his paper. Indeed, Rovelli proceeds many steps down the road of deriving the traditional formalism of quantum mechanics from basic information theoretic premises. (p 9 et seq).

---

8 We are always prattling on about "meaning". The discussions very often get disorganized and hard to pin down. One way to make sense of this is to see that each system takes the "meaning" of each other system in terms of what interaction occurs as to itself when encountering the other system. Does System A get sustenance from plants? Among the "meaning" of the plants to the eater thereof is sustenance, continuance. We humans can map out diverse and complex sets of interactions, and thus "meanings". But at the base of each is the effect of one organized system on one or more others.

On a separate front, Erich Joos (2006) and others labor to help us conceive of the connection between quantum mechanics and the "classical" universe.

Thus at the deepest level we now can probe, the realm of quantum mechanics underlying the universe as we experience it, we are led to see correlation between "systems" as a fundamental mechanic of how systems relate to each other, *and also to see the fundamental mechanics of the universe in relational terms.* We later will see that these concepts translate into seeing the Universe as a computational process, accessible through information theoretic tools.

Stepping back for a moment, we have looked up and out, and also looked down and in. In both the directions, we see, I suggest, that the central player in the organization, expression, indeed the creation of our tangible universe, is the mechanic of correlation. When one system encounters an element of another, there is a non-random likelihood that the "other" has a related component to be encountered also.

And thus I have ventured to suggest to you that this is the wellspring of tangible order in all its manifestations. We shall see in subsequent chapters that in the playing out of the mechanics of correlation is the history of the universe as we perceive it, the source of the "arrow of time", the foundation of causation, and much more.

Again, repetitively but necessarily, I note that in focusing on this mechanism, or phenomenon, I do not wish to imply that other aspects of the way the universe operates are not intimately involved in the formation and expression of correlated states of matter, or "order".

Obviously there are fundamental invariances which we identify in various ways. Among these are the Planck unit of action, the invariance in the propagation of electromagnetic fields, which we associate with the concept of the speed of light, the regularities in the processing of states of matter, and their relationships, which we associate with invariances such as the conservation of energy, the "forces" involved in integrating hadrons (though these "forces" can be seen as being involved in the diminution of degrees of freedom in relationships among hadrons and their constituents), the duality between particles and waves, and the like.

The many aspects of the Universe's operations which create "matter" and mediate its interactions are wonderful, and their exploration and expression in the undertakings of the sciences are not less than a magnificent set of endeavors.

This essay addresses how these aspects of what we call the universe create, or mediate, or manifest the non-random, integrated states which are, and define, the "order" which is us, and which is what we can identify as other expressions of the Universal mechanism -- in terms of tangible phenomena with which we can interact.

This is obviously not a complete description of all aspects of "ordering". But it highlights a realm of "order" and "ordering" which sheds light on our situation and experiences.

As we proceed from this point, we will find that the basic insights which Rovelli, among other quantum physicists, has put forward, shape, and simplify, understandings from all levels up from the quantum level.

Among those insights are that the universe is built upon and reflects probabilities at all levels, that reduced variance states, or correlated states, express order at all levels, that differentiated states "measure" each other, and form "relationships" at all levels, and that the "meaning" of any state lies in its relationships with other differentiated states, at all levels.

In my personal journey through his work, and the work at of eminent investigators here reflected, this connection, and correspondences, between the quantum level and the "macro", or correlated state levels, was at first a matter of conjecture, which upon continued attention hardened into an illuminating conviction. Now let us take a necessary next logical step.

2 Correlation=Differentiation=Order

The preceding chapter, identifies the correlation process, or reduction of degrees of freedom as between or among identifiable "systems" or phenomena, as the central element in the evolution of order in this universe.

This chapter addresses how this process plays out in the processes we can call "differentiation". This will lead to the concepts of "phases" of materials, and "phase changes" or "phase transitions", between them. This will lead to conceiving of "phases" as "relational regimes".

This terminology constitutes a simple and convenient conceptual step. It is not at all foreign to our thinking. But it will, if used broadly and consistently, help clarify a number of issues which have been fuzzy in much contemporaneous thought.

Close attention to these concepts, or ways of tracking order or organization in the Universe, will show how they are fundamental building blocks not only of order, but, naturally and necessarily, of our ways of perceiving and acting in the universe, to the extent we think accurately and usefully, and act effectively.

So let us proceed to setting up and defining a concept, "differentiation", in an architectural way -- in terms of this basic idea of correlations between elements. Let us say that if a set of elements relate to each other, by means of reduction in degrees of freedom between and among them, in ways differing from the relations which those elements would (were the elements not so correlated among themselves) have to other elements not so correlated, the set of correlated elements is "differentiated".

This attempt to state differentiation carefully leads to a somewhat dense paragraph. But illustrations from the world around us can, after we take one more step, make it seem normal and intelligible.

Let us take familiar example to make things simpler. Suppose one had a group of people milling around on a beach. For some – for the moment, any -- reason some of them link hands. Let us suppose that they do this in a simple and repetitive way, and this makes what we are familiar with as a circle.

Now, in this simple and repetitive hand linking, any given person in the circle has no hand free to grasp or be grasped by a passerby. No passersby can get into the circle, or attach to it (by handclasp).

Moreover, if this "circle" is to move – that is, alter relationships with anything outside it -- everyone in the circle has to move with everyone else.

Now, if the handclaspers move as a group -- and thus to maintain the group "identity", please note -- this group will have different impacts on other beachgoers than would movement by its individual person-components.

Miraculously, just by clasping hands, we have made a – differentiated -- "thing".

Suppose a unit of sodium chloride (salt), floating free in a liquid, locks onto a crystalline array – as when we grow salt crystal structures out of a solution of water and salt. We have here an act of differentiation.

But this individual phase shift, or relational shift, of one unit to others does not illustrate all we want to see. When we have a group of elements clicking in (or out) of a uniform or regular form of relationship, more is apparent to us. Let us therefore go in this direction for our example.

We are familiar with gases. We customarily describe a "gas" as a "phase" of matter. However, this is a little tricky. We conceive of a gas as a set of elements with elements moving randomly vis a vis each other. The elements of which have degrees of freedom such that they do not correlate with each other.

In this picture, we see no differentiations made up of groupings of the elements. We can speak of a "relational regime", if we wish. But it is a regime without regularities as among the elements vis-à-vis other elements in it, without reductions in variations among elements in it.

Now let us introduce correlations between some of the elements in this gas. Our example can be h^2o molecules grouping together and making a raindrop.

Degrees of freedom between the h^2o molecules are reduced. The molecules are densely associated with each other, but retain variability as to radial relationships, or angles. In our everyday, inherited language, the molecules slip and slide over each other, but stay close to each other.

We call this raindrop a liquid. We also call the "liquid" state a "phase" of matter. There is a distinct form, or way, of how the h^2o molecules relate to each other. The mode of relating to each other is uniform throughout the raindrop – the phase, or relational set, or field, or system.

We know a lot about liquids. We should, because there is where we life forms got our start.

One simple thing we have noticed is that the liquid state can be formed of many different elements and compounds – water, iron, salt, atoms as heavy as mercury, even very large things like ball bearings. Whatever the individual element, the set of relationships between them is what makes a "phase" – or, in the terms of this treatment, a system of relationships.

That is, liquids are made up of many different elements and compounds. But the way they act relative to each other, so to speak, is the same. And the way the liquid state relates to other distinct states of relationships is – roughly – the same.

We are so familiar with liquids, and solids – our extremely ingenious physicists and chemists have developed many, many treatments of these relational states -- that we need to pause and look deeper at several fundamental things which have happened here. They are things upon which our conceptual structures and language are based.

In this – liquid -- set of relationships, differing from the dancing gaseous variability outside it, we have the basics of differentiation of some "thing" from other things – we have, indeed, the creation of a "thing", in our language and thinking. As with the beach boys, or girls, there is an "inside" and an "outside". There is a "boundary" between the inside and the outside – between the set of relationships here assembled and other, differing sets of relationships – like "gas" and "solids".

The elements in the system are constrained. The group of elements taken as a whole -- the beach circle or the raindrop -- has relationships with things outside it which are different than would be the relationship of individual elements within the set to things outside the group were those individual elements interacting separately with the outside elements.

We take one more familiar step now, to "solids". Solids, like liquids, can be made up of lots of different things. What makes a solid a solid is simply further loss of degrees of freedom among the elements in the system.[9]

The result is that we have fixed "angles", and "shape". Symmetries in the way the universe works are now exhibited in "symmetric shape". More can obviously be said, but let us move on now.

Let us pause here to review and focus on how we are geared to the universe of correlations, *in terms of probabilities*, which, we remember, enter into all realizations of "matter".

In a gas, a set of randomly varying elements, we have no substantial likelihood that we will find another element right alongside one just encountered. And in most gases, except those highly concentrated, as in a sun, or volcano, or man-made jet, our senses do not register anything.

But in a liquid or solid, the probability of our encountering another component next to any given one is high. Our senses do register this composite event. And what we "see", or register -- drawing on the quantum mechanical perspective of hadrons as matter which excludes other matter, the quantum mechanical correlation perspective of Rovelli, and our biologically evolved systems -- is in effect the correlation.

In a liquid, just where we find a given h^2o molecule, on a set of x,y axes, is not determined, and we can move one around if we try to poke it. But in a solid, the next atom, or molecule, is determined in a way which we can express in our familiar form of x,y axes, or dimensions.

Degrees of freedom have been reduced. In this relational system, or phase of matter, if one set of atoms pokes at a part of the solid, it encounters resistance. The handclasp between the atoms, or other set of atoms, is set and firm (To a degree. Inject enough energy, or momentum, and you can get some shifting in the relationship, or disruption of it.)

Going back to the liquid, let us say now a river, we do not see the individual h^2o atoms. What we do see is a pattern of correlations which presents to us one object – the flowing stream of water atoms.

---

9 One can translate the differences in the relational regimes in terms of what Oscar the Observer observes -- or engages with. When Oscar encounters an element in a gas, there is no high likelihood that Oscar also "sees" or will encounter any other element in the gas, contemporaneously. When Oscar encounters, or sees, an element in a liquid, there is a high likelihood another element will be nearby, but the exact angle is not regular, or specified. When Oscar sees an element in a solid, the likelihood of encounters with elements nearby -- with little space separation or distance -- will be high, and the angles will be regularized, or non random.

With all this in hand, we can go back to an old puzzle about whether we can step into the same stream (usually thought of as water) or not.

We can indeed step in the same "stream" two or more times. (Actually, we know this until we start to confuse ourselves about the elements in the stream.) What we do is to step into the correlative structure -- the correlations. When we step into a stream – which is made up of the same types of elements with the same liquid relationships -- twice, we encounter the stream – the system of correlations -- twice, but we do not encounter the same atoms within that correlative structure twice.

One can focus further on this entity-defining system. When an element in a reduced-variance system of elements is so engaged, or entrained, or related, the element in the relationship is limited in its interaction with entities or elements outside the system in which it is engaged.

In the human context, one may touch a metal knife, let us say, with a finger. The result, as to the knife, may be to part one's own skin rather than extract metal atoms from the blade.

Not all encounters divide out results so sharply, but the basic mechanic is clear. Later we can discuss this in more detail as it relates to the human organization of groups. This fundamental characteristic is why we have human groups, and why we then have to go to the trouble of integrating those groups, or establishing relationships between the groups.

If we tarry to concentrate on these fundamental acts of order creation in the universe, we see the basics of concepts of "emergence", which are often used very loosely in scholarship and in ordinary discourse. The combination of the elements – in a relational regime or field, let us keep in mind – has created a "whole" differing from "the sum of the parts". This is the central aspect of "emergence" theory.

Note that the relationships of this "whole" – let's say a rock, or a cinderblock -- with other "wholes" did not exist before this relational set was constituted. A new set of relationships has been created. We tend to say one or more "characteristics" -- have "emerged".

What has "emerged" – let us keep in mind -- is the relationships of this composed regime (let's say the rock, or cinderblock) with other differentiation-created relational regimes (like a head, or other cinderblocks). The "characteristics" we speak of are the way the cinderblock impinges upon, or more generally relates to, the head, or the other cinderblock.

As we shall see in the later treatments of emergence theory and hierarchy theory, much of the confusion in "emergence" discussions results from not distinguishing between the two things briefly set out above – the composition of the differentiated relational field, or regime (a set of linkages of a given type), on the one hand, and on the other hand the resultant field of interactions between our chosen object - this set of relationships -- as a differentiated "thing" -- with other things outside it.

Let's, for exercise, elaborate on the basics of this order-creation by differentiation with our familiar standby, the liquid.

We talk about a liquid having characteristics. We unconsciously, often, shorthand what is going on when we do so. Let us take some of the "characteristics" of a liquid.

A liquid is often seen as "incompressible". If a "force" (let us say a displacement impulse) is put on it – let us say a lever in a vehicle's braking system -- that "force" will be felt – that is, will be exerted upon – all other relational sets surrounding and constraining the liquid. If the liquid is surrounded by ("solid") pipes except in an area where the liquid can move a solid structure (let us say a brake drum) that "force" can move, or displace, the brake drum. We have developed hydraulic engineering to exploit this "characteristic" of liquids.

A key feature is "incompressibility", which "means" (to us who experience the liquid) that when we try to move some molecules, and all the molecules as a group are bounded by some composite, we have to move that other composite, or we find we cannot move the part of the liquid which we try to move.

Let's take "buoyancy". This illustrates the interaction of more than one type of relational system, and "forces" acting upon them.

If we have a gravitational field and within it a liquid, constrained on all sides, but there are enough degrees of freedom in the liquid to move around vis a vis each other, and if we put some differentiated "solid" object on the liquid (an object which will maintain its own relational framework instead of merging into the liquid state with our chosen liquid), we see that if the solid object has more mass per unit of area than the liquid, the liquid will be moved out of the way of the solid and the solid will move through the liquid toward the center of mass attracting it until it reaches the center of mass of the system, or another solid – that is, a relationally bound system which it will not break up or move aside.

But if the solid has less mass than the liquid would have in the area where the solid abuts the liquid, the liquid will hold together and the solid will "float" on the liquid instead of ploughing all the way through it.

The boat will float. That is simple for us. But notice how many circumstances we had to specify – or to assume – to get to what we call simple.

We could illustrate some equivalent and some differing external relational effects with "solids". But let us move on.

The basic point to be illustrated is that the universe we experience is made up of fields of relationships -- relational systems -- which are differentiated from other sets of relationships. Our internal maps of this universe track the interactions between these differentiated relational state set, or structures. Every differentiated relational field exists – that is, is "measured" by, has an operative set of characteristics -- in relationship with other differentiated systems of relationships.[10]

And, to parallel Rovelli's statements, that is a complete description of the (visible to us) universe.

---

[10] As the next section elaborates, this is necessarily and unavoidably the core of "emergence theory".

Now let us focus again on how fundamental is this process of correlation-> differentiation.

Again, let us look at the night sky, or a picture of it. The stars are differentiated from the void in that they are reduced-variance composites. We conceive of stars as made up of gases. But the key to focus on, when we star-gaze, is that they are clumps of gas, differentiated by having a large volume or set of elements staying closer to each other than to elements in the "void" "outside." The stars have the basics we have mentioned – an inside, a boundary, an outside, and a composite set of relationships with other or groups of stars.

Let us go to galaxies – with the assistance of pictures, since it is a little hard for us, using just our eyes, inside our galaxy, to see a galaxy and its relationships with other galaxies. Our galaxy can be pictured as a swirl-pattern of stars. These stars are more bound to, or correlated with, each other than to/with stars outside the galaxy. (Globular galaxies look more like big round clusters, instead of swirls).

For practical astronomical purposes, so to speak, this differentiation defines galaxies. And so all the way up the astronomic hierarchy. The night sky is a vast tiering of differentiations. Equivalently, it is a vast tiering of correlations.

Let us go back to the atoms other than hydrogen atoms. We have seen in the prior chapter that composite atoms are made up of differentiated groupings of hadrons. Molecules are made up of differentiated groupings of atoms. All the vast complex of interactions between atoms, molecules, and their aggregates is based upon and resultant from this process of differentiation.

Looking over this vista, we can assay another broad observation. All the situations in which "work" (in human terms) is done (where we have energy differentials, chemical differentials, etc.) result from differentiation, resultant from aggregation, resultant from correlation.

This is frequently, and fundamentally, overlooked in discussions of what we call "energy" and where it comes from. For example, we speak of the "potential energy" available from dropping a weight from a height, or energy from the Sun's radiation.

But the height (on earth) was created by the gravitational attraction bringing together elements, and the "potential energy" expresses the differential on the Earth's surface between one position in the gravitational field and another.

The sun is composed by gravitational attraction, and the "energy" streaming out of the Sun consists of a means of dissipating the differentials created in the sun's integration of the elements attracted into it.

Thus our visible, tangible universe has been built.

This is not complicated. This simplicity does, though, create complexity, as will be explained in a later section.

For the moment, we can observe that differentiation occurs in a great many ways, given the variety of atomic structures and the ways they can and do come into close relationship.

In addition to the other examples here provided, for an illustration of how, over the ongoing evolution of the Universe, differentiation creates distinguishable sets of relationships, let us look at the picture our scientists have constructed of the earth on which we stand. (One is tempted to say "our" Earth, but we belong to the Earth far more than the Earth belongs to us.)

Over billions of years the restless shifting of atoms of different weights and electronic characteristics in the Earth's gravitational field, plus the heat resultant from accretion of the huge mass of the Earth, shifting of the heavier elements toward the core, and radioactivity, has sifted out distinct relational regimes, as is illustrated by this depiction.

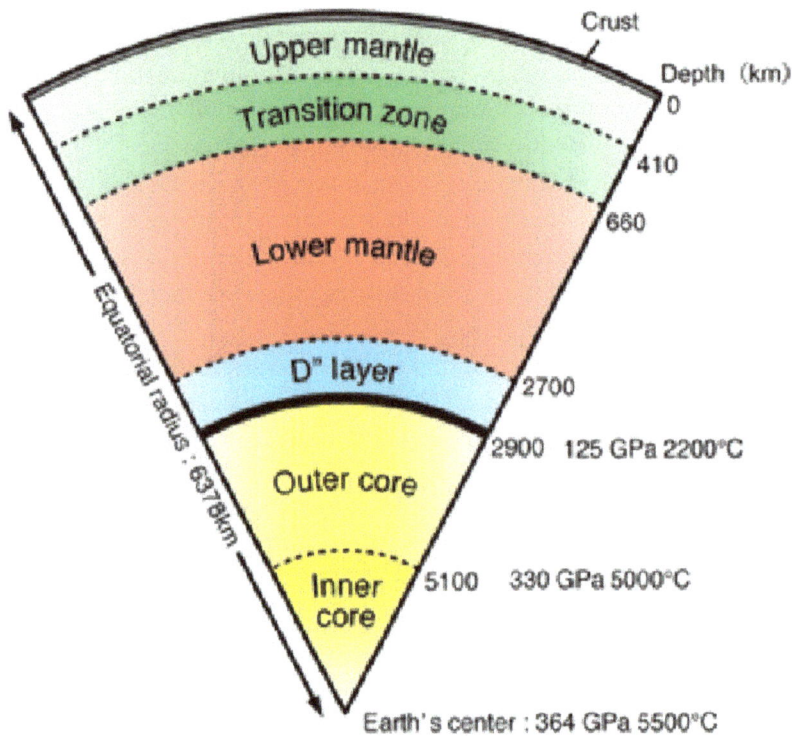

Fig.2.1: Some of Earth's Distinct Relational Regimes.

Other planets in our solar system are also depicted as having layered, differentiated, interiors, though the particulars vary, largely resultant from sizes of the planets, and distances from the sun.

The building of this universe in this simple way has at times seemed inconsistent with some understandings of the "second law of thermodynamics". People have long noted the contradiction between a universe apparently steadily cranking out order and the predictions of eventual "heat death" from the equilibrating processes implicit in the Second Law.

We see that adding "energy" to a system of interacting elements can lead to breaking of correlations in it: letting the system "cool" leads to correlational phase changes and order.

The Universe is now depicted as rapidly expanding and cooling. Therefore correlations and phase transitions to differentiated, ordered states can and do occur in this Universe. (Actually, they make up -- *they are* --the tangible universe which we, differentiated correlational states, register in our naive, basic organic state).

How does this evolution of differentiated, ordered, states fit into the cosmological story, which so many have suggested tends to disorder and a "heat death"?

Two prominent theorists, David Layzer (1990) and Eric Chaisson (2001), offer a picture consistent with the view of a correlational, ordering universe. They suggest that we look to the expansion of the Universe to frame an understanding.

In rough terms, they suggest, the Universe's rate of expansion allows more opportunity for variation within it than can be realized, or actualized, on a current basis as the expansion proceeds. Thus the disequilibrium, or reduction in the extent to which complete randomness is realized -- which we here call order -- can arise and persist.

However, one might also address the arguments expressing the ubiquity of the equilibrating effects expressed in the "second law" by noting that all this equilibration proceeds from the differentials created by correlation. Doing this seems to turn the story on its head.

Thus are intertwined the "expansion" of the universe and order in the universe – indeed, the existence of the universe itself as we are able thus far to depict and interact with the "matter" in it.

What we see as "expansion" has brought our universe into being. But from the viewpoint of a manifestation of order, ourselves, might we not be tempted to suppose that the efflorescence of phase transitions which has allowed correlation and differentiation to be expressed might itself be a primary driving factor, with expansion being a corollary?

Though this suggestion is tempting, I cannot establish this connection at this time, and rest with the Layzer/Chaisson explanations for now.

Many have wondered at and about the concept of an "anthropocentric" universe – how could it be that the vast machineries of the Universe are tuned so as to produce us?

That question begs for an analogous question from the relational point of view. Here it is. If the order-generating process of this universe has produced us, the question arises – what are the characteristics of the correlational process itself, taken with other constants in the Universe, such that correlational regimes, can eventuate sufficient complexity to produce living things like us? From this perspective, trying to look at correlation from inside out, so to speak, why can this correlation/differentiation process proceed as it has?

At this time, researchers are actively probing the characteristics we have come to call 'entanglement' and correlation at the most fundamental levels presently accessible to us. One may hope that their accumulated and refined insights may come to address this sort of question.

Lastly, as to "phases" of matter. I have restricted this discussion to phases, or relational regimes, very familiar to humans in our limited range of perceptions and actions. We are all familiar with gases, fluids, and solids. (I have not included plasmas, as this does not seem to connect well with ordered states.)

But the life process on this Earth, and humans' chemical inquiries, have produced and revealed a large number of differentiated sets of relationships, some much more intricate and elaborated than bulk phases like gases, liquids and solids. For just one of many examples, "liquid crystals" (now widely in human use) exhibit states intermediate between solids and liquids. See Thomas and Weber as to mesophases in liquid crystals, for examples which were thought exotic only a few decades ago, but now woven into industrial civilization.

One have "miscible" and "immiscible' liquid phases (liquids which will mix or not mix). The existence of long, flexible, strings of matter are woven into life (e.g. muscle fibers and DNA strands). Thus, this chapter serves only to set out a very basic conceptual framework into which a variety of relational systems can fit.

3 Some Basic Characteristics of this Ordering Process as Thus Conceived

The preceding two chapters have described the reduction of degrees of freedom between relatable systems, or phenomena, as the foundation of the creation of order, the ordering process of the universe, and that this process creates differentiations between relational regimes, or sets of systems of relationships.

This chapter sets out some basic characteristics of this ordering process in the universe, within this framework. Taking these characteristics into account makes this ordering process more intelligible, over a broad range of manifestations.

Here I summarize some apparent characteristics briefly, and then expand upon them to some extent.

Order construction, and hence the Universe's construction, is

1. Combinatorial;
2. Computational;
3. Hierarchical;
4. *Probabilistic as to each increment or decrement* – that is, the process as a whole proceeds, as it builds hierarchies, as a stepwise realization of conditional probabilities;
5. Subject to fundamental statistical regularities at all levels – that is, power law distributions, or curves, and entropy/order formulations of the Boltzmann-Gibbs-Shannon and derivative types;
6. Operative at each level of aggregation, and across levels of aggregation; and
7. Operative in "dynamic" as well as "static" conditions.

Let us take these characteristics in sequence.

3.1 Order is Combinatorial

Let us start with the combinatorial aspect of order building. The reduction in degrees of freedom among elements, or systems, creates non-variant relationships, which constitute what we perceive as and call combinations. The combinatorial nature of order construction proceeds – as I hope is now tautologically evident, immediately and directly -- from the fact that elements must be combined, or linked, to make aggregates (and then those aggregates must be combined to produce the next level of aggregates, etc.)[11]

---

11 There is, in this paragraph and in others dealing with the fundamentals of order construction, something of a tautological element. We are defining a system in terms which are themselves created by its construction. See Godel if you want to grapple with this more.

In a sense, we have to rise above this tautological basic, or just accept it, to see the power of the observation.

One demonstration of that power is the English mathematician George Boole's enormous contribution in constructing universally powerful logic systems simply by specifying the "and" relationships, and the limits of them by the "not" and "or" operators.

We fortunate 21st century users of the internet and its search engines, and related search algorithms, routinely and seemingly effortlessly use the power of Boole's crystallization of insights. We simply string together, or associate, words and evoke the vast field of combinations found in an enormous range of codified materials. (Google exploited this system by also developing aspects of relationship construction.)

The ways in which organization is combinatorial are coterminous with organization itself. The basic point to keep in mind is that always and everywhere, we are dealing with combinations. The more combinations which can be expressed, the more order is elaborated.

3.2 Order is Computational

In recent decades, leaders in physics, mathematics, and related fields have pointed to, as I would characterize it, an underlying identity as to the physical universe, information, and computation. Among those prominently associated with this line of thought are John Wheeler, Stephen Hawking, Carlo Rovelli, Jacob Beckenstein, Seth Lloyd, and, in a way, Max Tegmark. (The brevity of this listing undervalues, I fear, the great many able investigators who are well aware of the concepts and use them productively.) I have selected from this list some to focus upon.

As I pointed out in chapter 2, Rovelli puts information theory at the heart of quantum mechanics.

The computational nature of the order generation process is maintained by, or expressed according to, invariances such as the speed of light, the Planck constant, "conservation of energy", etc.

By stating that order is "computational" I do not intend, obviously, to suggest that order in the Universe or the ordering process of the Universe is or are structured along the lines of current computing machinery, the Babbage machine, or the application of chalk to blackboards (to get really antique about the discussion).

Rather, I merely suggest that order has discrete fundamental factors and a set of invariances which are integral to all its manifestations. This allows us, its creatures, to proceed along what we call computational lines.

I fear that this emphasis on fundamentals and use of language which is not widely familiar at this time may lead to some brow-furrowing as to the sweep of the proposition. Also, possibly there may be impatience as to lack of specifics to support and illustrate it. But the basic proposition that order generation is computational and can be analyzed by computational means should not be controversial. Perhaps enough is said for the moment. After all, order is, and could not exist without, regularities underlying our encounters with it. And those regularities make possible computation.

3.3 Order is Hierarchical

The first chapter sets out fundamental factual outlines supporting this proposition. The second chapter outlines the supporting dynamic.

Since order necessarily must be built upon the process of correlation, as each correlation is built into an entity, other correlations may be built between this entity and others.

The simplicity and exclusivity of this dynamic accounts for the ubiquity of hierarchies in the universe, and thus of course in our own experience. When you correlate correlations, you have hierarchy.

We observe and participate in hierarchical situations pervasively. But we have invented many situational ways of characterizing our experience. This has led to attempts to construct hierarchy theories in greater than necessary abundance. More about this in the chapter on hierarchy theories.

3.4 Order, and Thus the Sensible Universe, is Built as a Continuing, Parallel Cascade of Realized Conditional Probabilities

This proposition may justify a little more pause to take in.

At the quantum level, we have come to see Schroedinger's wave function as an expression of the probability of making an observation as to the presence (or correlation) of a system at a given point and time. We speak as to quantum observables, or expression of "matter", in terms of probability density matrices. We make probability maps of electron observables, or the interactions of what we call electrons with other quantum systems (conceiving all systems as quantum systems).

Thus we can say that at the most fundamental level we have thus far managed to reach, we encounter the universe, and thus must conceive that it is built, *in probabilistic terms.* We can see that all hadron, or nucleon, related realizations of the Universe's potentials, at least, occur with a probabilistic element.

To illustrate, let us imagine the career of helium and beryllium atoms which are fused, in a star, into a carbon atom. Only some, not all, such helium and beryllium atoms get fused in the star. This is a probabilistic matter. And a low one. But let's have these lucky light atoms become a somewhat more hefty carbon atom.

Whether that carbon atom were later, after being blasted out of the star, to be involved in, let us say, a condensed matter aggregate, such as a planet, would also be a probabilistic matter.

And if that carbon atom would get involved in a life unit, that would be another low probability event, considered from the point of view, had it a point of view, of that carbon atom.

And so forth.

Now let us conceive the events of stars, planets, and living units on planets as *a tiering of probabilities*. That is, we have a probability on top of a probability on top of a probability, etc.

Let us recall the proposition I have advanced thus far -- that what we can consider to be ordered relationships map one to one on the visible universe. So now what we perceive to be matter and its manifestations is expressive of, or more accurately embodiment of, correlational events yielding correlational structures.

If this is so, then the entire tangible universe, and all in it, can be considered a structure of realized conditional probabilities.

Take a simple skein -- if suns, then complex, heavy atoms; if complex, heavy atoms, then planets with differentiated elements, and differentiated structures; if such planets, then water on some of them; if water on some, then life on some of those having water; if single celled life on some planets having water, then multicellular life on some planets; if multicellular life then complex multicellular life, on some such planets; if complex multicellular life on some such planets, then complex social systems and "civilizations" on some of these planets.

Frank Drake did precisely this in setting out his famous formula to approach estimating probabilities of life in the Universe, available to the general readership at https://en.wikipedia.org/wiki/Drake_equation. The formula is just a tiering (or, if you will, a nesting) of probabilities.

Considering the universe both as a tiered realization of order and as as a tiering of conditional probabilities has interesting implications.

First, we see this traced out in the distribution of masses and numbers of complex atoms. Larger and more complex are rarer than simple and less complex atoms.

More complex assemblages of complex and simple atoms (read more complex relational regimes) are less common than the simpler systems on which they are probabilistically built. Life, a complex pattern of dynamic processes, is rarer than the complexes of which it is made. More complex life is rarer still. And so forth.

As noted before, Eric Chaisson has done a beautiful job of quantifying the masses of matter involved in things like galaxies, stars, planets, living organisms, etc. More, he has quantified the rates of energy flow through such structures. And this has led directly to measuring the complexity of the tiers of organization which he has identified and measured.

Chaisson's primary tool of measure is the "energy rate density" of a considered system -- that is, the amount of energy flowing through the systems per unit of mass and time. In chapter eleven I propose that the higher energy rate density the systems, the rarer in the Universe they will be. This is necessarily so if we see these states as results of tiered correlational events.

Chaisson seems to project an increasing level of organization, and energy rate densities, over time in the Universe. That is, over 13 billion of years of cosmic evolution, tiered correlation building has funneled more energy into high energy density, highly ordered dynamics, such as us and our artifacts.

Clearly our local situation on Earth demonstrates that over a few billion years in this solar system a tiny, and calculable, fraction of the mass has evolved into relatively high energy density, and thus complex, phenomena. This is documented later in this manuscript.

Presumably this has been replicated, in system terms if not in the detailed particulars of each and every manifestation, in other parts of the vast and evolving Universe.

But the question arises -- how long and how far will this order building go, if we try to project it across the -- not as yet fully defined -- scope of the Universe and its processes?

Will there be an equilibrium, or steady state, or asymptotically realized, conditional probability realization, and over all level of order realization in the Universe as it evolves?

Others are better situated to address this issue in a quantitative way than am I at this time.

The researchers into computation and the "holographic universe" seem to suggest more "entropy" (conceived as component arrangements) with increasing bounds on the universe. Layzer and Chaisson seem to suggest that order can arise with the expansion of the universe. This would seem to imply that the more expansion, the more order. One might express this in formula terms.

So far, all seems to be consistent. But if there is a constant slope to a complexity realization curve in the universe, as I do suggest later, one is led to ask what mechanism would raise the total complexity level in whatever the Universe is, taken as a limited and defined whole, and whether there are boundaries on that mechanism.

3.5 Order is Earmarked by Ubiquitous Power Law and Boltzmann-Gibbs-Shannon Entropy/Order Statistics

Mark Buchanan (2000), in his book "Ubiquity", provided considerable documentation to show that "power law" expressions[12] crop up throughout nature. Hence the title of the book. We see "power law" distributions of events in earthquake sizes, city sizes, war sizes, planet sizes, the abundance of elements, the statistics which emerge from "phase transitions", etc.

---

12 Very roughly speaking, "power law" distributions appear as straight lines on graph paper using logarithmic scales. Take earthquakes, as a convenient example. Lots of little tremors, the frequency falling off by powers of ten, as usually scaled, as violence increases by powers of ten.

Reports of power law sitings continually emerge, in various settings. It seems we discover and rediscover this phenomenon wherever we look, often without seeming to be aware of the universality which Buchanan pointed out years ago.

Alberto Laslo Barabasi, (2002) echoed Buchanan's theme, which he had been pursuing himself for some time, and, importantly, refined it as follows

> "In ordinary systems all quantities follow bell curves, and correlations decay rapidly, obeying exponential laws. But all that changes if the system is forced to undergo a phase transition. Then power laws emerge – nature's unmistakable sign that chaos is departing in favor of order. The theory of phase transitions told us loud and clear that the road from disorder to order is maintained by the powerful forces of self-organization and is paved by power laws."

In other, and equally global, words, correlational processes generate power laws.

Barabasi suggests that this may result from "preferential attachment" processes -- relationships accumulate by relating to existing relationships. The rich get richer. One can think if this, roughly, in terms of a larger set of correlations having a basis for more additional correlations than a smaller set.

Both the logic and the evidence seem inescapable -- a universe built by correlational processes must have the power law signature all over it. That goes from foundations up, as the comments on information theory and quantum mechanics in a preceding chapter indicate.

3.6 Correlational – Relational – Dynamics and "Forces" Operate at Each Level of Aggregation, and Across Levels of Aggregation

Nucleons can be co-located, correlated with other nucleons. That is the way we get complex atoms. Complex atoms can form relationships, or bonds, with simple atoms -- as in carbohydrates. Many atoms of different sizes and shapes can be correlated so as to make molecules. Molecules aggregate with other molecules. We can have "solid" systems – e.g. small bits of solid matter – floating in gaseous systems, like the atmosphere; liquid systems encased in "solid" systems – e.g. brake fluid in an automobile chassis. We can also have gases enclosed in solid systems – e.g. gases in a boiler, etc.

Living systems demonstrate intricate combinations of relational systems with differing degrees of freedom – e.g. bones, circulatory systems, lungs moving chemical elements between gaseous, liquid, and intermediate states.

As previously observed, at and across all levels and combinations of levels of states, combinatorial power laws appear.

3.7 Correlational Processes Operate in "Dynamic" as Well as "Static" Conditions

Let us first focus a moment on what we mean by "dynamic". Let us look under the covers of our language system.

We do not ordinarily use the word to refer to a hot gas with entirely random motions within it. Implied in the way we use the term is a situation in which there are some correlations within the system, but the correlated elements embody some motion (i.e. manifest some degrees of freedom) relative to other elements, and relative to an observer system.

Let us take a common example, the "Benard Cell". Given a thin layer of fluid with one external surface in contact with a hotter surface or environment than the other external surface, the fluid can, and under controlled conditions, will organize itself into loops which transfer motion – heat – from the hotter to the cooler sides of the film. These loops line up in a regular hexagonal array in the liquid.

Some, notably Eric Schneider (2005), have used this system as a part of a paradigmatic argument as to how order is created – i.e. from equilibrational processes in dissipating differentials.

Let us look at these cells from the viewpoint of correlation manifestations. There are several lessons to be drawn.

First, we do not focus on, or see, or register, the individual movements of the individual water molecules (as with the river example). What appears to us is the pattern of correlations. Though individual water molecules move in a circular flow, at any given time, or point, what we see is a set of molecules in a given fixed relationship to adjoining molecules. (This has an analogue at the quantum level.)

Secondly, this regularity can be seen as an energy minimization phenomenon, just as we can see as energy minimizing phenomena phase transitions from gaseous to liquid relational systems and from liquid to solid systems.

Thirdly, in a section on "emergence", I will point out how highly complex dynamic processes can be related to conceptions of "top down" "emergent" effects.

Other lessons can be drawn, but we can defer them for now. Let us go to other examples.

The great "red spot" of Jupiter is a convenient illustration. In the gaseous boundary area of Jupiter is a large (enough to hold lots of Earth size planets) counterclockwise swirl of gaseous matter. It is clearly differentiated from nearby streams of gaseous matter, and has been for at least hundreds of years.

Again, what we "see", or natively register, is the correlation – the persistently maintained swirl pattern of relationships between component entities – rather than the individual elements of which it is composed. We might refocus on individual components of the elements in that swirl, but the "Red Spot" is the pattern – the regularity in relationships which is manifest to an observer system.

One could also observe that we living creatures are persistently maintained combinations of dynamic patterns. In our interactions with each other, we do not usually register the individual components (e.g. the calcium atoms in the bones, the oxygen atoms in the blood cells, the cells themselves.) We exist as a dynamic mechanic. We are a dynamic pattern of flows of energy and materials which manifests, on a continuous basis, both degrees of freedom -- motions as between elements -- and correlations between elements,

When boxers and football players enact their violent conflictual dynamics, they pound intricate dynamic structures of fist, or shoulder, or whole body, each against the dynamic structure of the other. They seek to displace, or to disorganize (within limits, in civilized societies) the dynamic patterns each of the other.

So with armies. On the large maps consulted by generals, and professional imaging organizations like the Press, we see the patterns of the dynamic organization which is the army -- the large scale space occupying effects, the displacement of the other army, the disruption which each army does or does not cause to the other.

Enlarging the scale, we can depict, as does Eric Chaisson, the Universe as embodying dynamics. We see the swirling which makes up galaxies, the influx of gases into stars and radiation from the stars, the creation and disruption of star, planet, comet, and other aggregates, on grand scale.

We see a progression of states in the Universe, through the early phase transitions which made up matter, through a stepwise process of atom building in stars as complex elements were formed, through the cosmic structures we can now see, with our augmented senses. We also imagine, hazily and tentatively, different cosmic scale configurations of our universe in far-future epochs.

And the universal system both creates and disrupts correlations -- patterns, states of being -- as it grinds toward unknown eventuation.

This much, I suggest, we transient bits of pattern realization can easily see. How to mesh our own intricate computation/navigation within the Grand System which made us is our constant challenge.

4 Hierarchy and Emergence Theories

The previous chapter suggests that the universe is playing out correlational potentials in an *hierarchical, probabilistic* fashion. This simple picture provides secure physical grounding for both "emergence" theories and hierarchy theories. In doing so it harmonizes them, as aspects of a common theory of order development.

Hierarchical depictions of the development of the universe abound at this time. They are often idiosyncratic in classifications and categories. They often seem to convey the gist that over time, or the process of the universe's evolution, atoms combine into various categories of manifestations, and we arrive today at some complex creatures doing complex things. I particularly like the illustrations on p. 4 and pp. 149-208 of the 2001 edition of Eric Chaisson's "Cosmic Evolution." (Chaisson is of course much more detailed and fact-supported than this brief paragraph.)

The preceding chapters suggest looking in two directions to see hierarchical development, based on correlational mechanics – the galactic scale on the one hand, and on the other the composition of complex atoms, and then of combinations of those atoms in layer-like fashion.

Both directions reveal a continuing process of correlation development, though the particle-up, so to speak, evolution may more obviously suggest a direction of time than the galactic ensembles picture does. (We can reconcile these two perspectives into one by seeing the cosmic array as the beginning, so to speak, or substrate, if the atom-up construction. We see correlation in two "directions" because we are intermediate in scale between the atom on the one hand and the cosmos and the other.)

The picture here thus far presented has features which are often found in hierarchy theories which attempt some clear connection of theory with physical structures. Among these common features are the ideas that there are "levels" of aggregation, that parts make up compositions, that one may have different descriptions of the interactions among the parts than the descriptions of the behavior of the composed entity with respect to entities other than the composed entity, and that the parts often may interact with each other more rapidly than does the whole interact with other things. One can see perceptions such as this in Salthe (1985), Ellis (2006), and Corning (2002), among others.

Ellis also has observed that a given type of "whole" may have differing parts, or elements – as in a liquid state made up of different types of atoms. This is implicit in the relational regime approach used here.

So far, so good. We are writing a common playbook with many common paragraphs. The previous chapters of this book go beyond most hierarchy theories I have found to date in

1. characterizing combinations of elements in terms of regimes of relationships,
2. insisting that each level of aggregation will itself potentially engage in one or more relational regimes external to it (at least up to the level of the entirety of the Universe),

3. pointing out that the characterization of the behavior of the "whole", the composite system, will depend on the relational regime(s) in which the "whole" participates,[13]
4. bringing to the fore the realization that correlational mechanics operate at all levels of aggregation, and correlation-based statistics apply at all levels and in all types of aggregation, and
5. featuring the realization that all aggregations – that is, all realizations of ordered states – can be seen as progressive realization of conditional probabilities.[14]

Putting these additional features into the hierarchy picture gives one bases for quantitative characterizations of different levels and types of aggregation, or, put another way, the statistical characterizations of the realization of hierarchy potentials.

In addition, the first three chapters
1. offer a perspective on dynamic phenomena from a relational regime standpoint,
2. point out that ordering -- that is, correlations (and hence hierarchies) -- can and does exist in dynamic processes, and
3. point up that such ordering can be attributed to energy minimization, as in more "static" circumstances.

When we add these features to hierarchy theories relating to observable physical systems, we come out with an hierarchy theory which has a one to one correspondence to the physical realization of the visible universe – i.e. the evolution of order in the universe.

Let me repeat that this manuscript focuses tightly on the evolution of the universe as a physical phenomenon. I do not here attempt to address the full range of terminologies applied to hierarchical organization in human representation systems – in our languages and graphs. Given that we exist in an hierarchically organized universe, one would expect that our representation systems would show hierarchical organization.

Also, I have not here attempted to sort through all the ways in which statements of hierarchy theories use designations of areas of academic study as markers in hierarchy theories.

---

13 To me there is a provocative and beautiful parallel between this insight at the "macro" or "classical" level and Rovelli's insights at the quantum mechanics level -- i.e. that each quantum system measures the other, and the two are defined in effect in their relationship. The parallel I suggest accords with the frequent assertion by Rovelli and others that all systems are quantum systems.

14 As noted before, here also is a parallel with the quantum level. The entire correlational mechanic at all levels may be said to play out in a field of what might be called randomness, or if you will, alternatively, only in a probabilistic way.

Nor does this approach seek to take up individual situations in which authors sometimes seem to use statements arising from representation system structure separately from direct reference to physical phenomena, and sometimes mix statements arising from representation system structure with statements which are keyed directly to physical processes.

All this proliferating, inconsistent terminology will eventually be sorted out. Ultimately we will be forced to get close mapping between the physical processes and the representation systems. I have therefore focused on the physical substrate for hierarchy theories, and have tried to keep the language offered closely tied to that physical substrate.

Now let us move to "emergence" theories. We have mapped hierarchy theory, at least as to physical systems, to the concepts here summarized as to the generation of order in the universe. This leads directly to mapping a physical basis for theories of emergence to both the concepts used in this manuscript up to this point and to hierarchy theory.

First, though, let me recognize that considerable – one is tempted to say enormous -- creativity has been exercised in grappling with perceptions of "emergence". The result, however, is a jumble of perspectives and terminologies. One is confronted with notions such as "strong" and "weak" emergence, "ontological physicalism", "property emergence", "irreducibility of emergence", "downward causation", the suggestion that "emergence" is associated only with complex systems, and much, much more.

If one were to attempt to follow each logic on its own terms and then to try to reconcile the perspectives using the terminologies found in them, one would set on the table for oneself an indigestible stew of creative but loose language, lacking conceptual consistency.

One might be tempted to create yet more neologisms and flights of conceptualization, heaping the jumble still higher. I would urge a reader to decline the temptation.

Writers such as those I have cited do, rather uniformly, recognize that emergence arises from combining parts into "wholes", and these "wholes" interact with their surroundings. That gives us a starting point in relating the literature on "emergence" to order development theory and hierarchy theory.

If we put emergence theory in the framework of physical realization of correlational potentials in an hierarchical fashion, conceptualizing "emergence" is immediately simplified. The simplification comes from thinking simply and directly in terms of combining elements into "wholes", these "wholes" *then exist in relation to other systems*, as Rovelli specified even at the quantum level, and the characteristics which the "wholes" then have are just the terms of the relationships between them and those systems with which they have interactions.

As Peter Corning has observed "...the debate about whether or not the whole can be predicted from the properties of the parts misses the point. Wholes produce unique combined effects, but many of these effects may be co-determined by the context and the interactions between the whole and its environment(s)."

Right on target.

If one then recognizes that this process is iterated in hierarchical fashion, one has the framework for keeping the conceptual books for "emergence".

To recapitulate, at bottom is the combinatorial process, or correlational process, or the relational regime, as to any given set of elements. These elements are the "parts" which when involved in a system of relationships make up the "whole".

We then look to see the relationships this "whole" may have with other elements and aggregates with which it may interact. These relationships of the "whole" with other systems define or constitute the "emergent" characteristics -- again, relative to other systems -- which have resulted from the combination of elements in the system on which we have focused our attention. [15]

We then look to the continuation of this pattern of pattern development, as with hierarchy theory. That is, atoms into complex atoms, atoms into molecules, into combinations of molecules, etc. At each step we have the recapitulation of the two steps above.

Taking an overview across space and the process of realization of the Universe, one way of stating the matter is that as the universe proceeds in this fashion, we have a continuing process of emergences. In another way of stating the process, we have the filling out of the set(s) of reduced variance, or ordered, states which make up the sensible universe.

As stated before, we are here merely stating in "emergence" language the evolution of the sensible universe in an hierarchical fashion.

---

15 Perhaps this repetition of the core assertion in the thesis of this book warrants an illustration. Let us provide it by stepping up a few levels of aggregation to use a familiar object, as suggested to me by Jerald Robertson, offering the example of a bridge. In the bus on the way from the airport to the ferry to the 2008 conference on Star Island, Jerry posed the question of what could be said concerning a bridge transported to outer space. Jerry suggested that little or nothing could be said about it.

We both immediately understood the assertion and did not dwell on it. Let's here elaborate on it a bit. Let us put the bridge in one of the great voids between super clusters. None of our instruments could register it. Its mass would be negligible with respect to any sun or cluster of suns. To a dust grain or small bit of rock, it could be a sudden stop, if the grain or pebble happened to hit a girder rather than an open space in the lattice work. To the rare photonic field effect, it could correlate with a transmutation into excitation of local atomic structure.

Now put the bridge on Earth, spanning a river. The bridge is the support for rolling vehicles, a weight pressing on its foundations, a path for pedestrians, a connection between, let us say, one State and another, a facilitator of trade flows, a shade for a homeless person seeking a place to sleep, the locus of periodic maintenance, an energy sink by day and energy radiator by night, should it fail an object of intense scrutiny of its design and the maintenance policies of relevant political jurisdictions -- and so on with hardly an end in sight.

The structure of a bridge is shaped by and for the complex set of dynamic processes – the complex set of relationships -- into which it is designed to fit. If it is not in that complex, the bridge is not a bridge. What else we might want to call it would depend on how it might be encountered.

This simplification, based on the correlation mechanism, is particularly useful in dealing with living creatures, the focus of much of our human attention. The very brief synopsis I offer here is spelled out in more detail in the "Life" chapter.

Let us look at where we have "energy flows", or differential minimization, which allow complex patterns to arise and to persist. We have thought in terms of chemistry at the base of this subset. Here, initially in the oceans of Earth we have seen the development of autocatalytic, cycling action patterns which can reproduce themselves.

Over time, such pattern reproducers have generated systems, or patterns, in the reproduction process itself which further elaborate state sets (forms, as we see them) over time. We call this biological evolution. In this evolutionary mechanism we see a fascinating interaction of order and randomness, which I will touch upon later.

In this elaboration of state sets in the reproduction chain, the systems of ordered dynamics which we call life creates hierarchies of its products. In Earth's ecosystem we see early-on single celled organisms, then multicellular organisms, thence groups (or societies) of multicellular organizations, and groups of groups of multicellular organisms. Life evolution makes groups, and groups of groups, just as the universe has done in the chain of aggregations which gave life its substrate.

Now let us deal with the issue of how we tend to perceive and to define emergence in the activities of a complex life system like a human.

We can say that a complex creature like humans can compose a large number of action sets, with which to interact with other humans and with other objects. We still have the compositional processes within the human, and then the collective effects, or relational interactions, as between the human and other humans, or other things. We just have a multitude of, or complex set of, interactions.

And we can, of course, have groupings of humans following the same iterative composition/emergence pattern up the tiering of organizational structures of humans.

To illustrate that the interactions of the composites which humans form of themselves could be simple or complex, let us take two armies (I have spent a fair amount of time on military history, so I tend to use military examples.) Each army composes action patterns, to be directed toward the other. "Emergent" characteristics of each army include its relationships to (and effects on) the opposing army, the opposing citizenry, and the environment.

If the armies were to engage in trench warfare, as in World War 1, they would not exhibit highly complex interactions. But give one or both armies high mobility and a variety of ways of impinging on the other, as in World War 11, and subsequent conflicts, and we get more complex action patterns like blitzkrieg, encircling and flanking tactics, etc.

In the 20th and 21st centuries thus far, warfare has been taken to the level of trying to break down the entire organizational structure which has created and sustains the opposing army. An "army" becomes a barrage of physical and electronic impingements disrupting the opposing social system in its most basic organizational functions -- food, transport, communication, etc.

Thus far, the logic I suggest would seem simple and direct. However, seeing the logic in action is not always so simple. Our position as individual elements in a compositional process making up aggregates like cities, states, and armies often leads us to confusing use of "emergence" terminology.

We can see ourselves combining with other elements in large scale action patterns, some of which are complex. It is natural for us to see this process as an "emerging" or "emergent" phenomenon. When we do this, we lose track of the two steps in composition and effect of composition. Having done this, we have tended to obscure the over all process of tiering of compositions.

Let us move on to some other elements of the complex stew of notions about "emergence".

As Corning pointed out, we sometimes tend to confuse the picture when we say that the properties of the "whole" cannot be predicted from the characteristics of the components.

To take an example, the field of relationships making up water droplet falling toward a river are merged into the relationships of the water molecules in the river, and the droplet loses its inside, outside, differentiation and identity. But we do know much about how the liquid in the river behaves.

The field of relationships in the droplet and in the river are not the same as the relationships of the river to the land. But we have also worked out descriptions of how rivers flow upon and affect land.

More generally, we have knowledge of how liquids interact with solids and gases in various situations. We have a good idea of how various combinations of iron atoms and "additives", when cast in solid form, interact with other composites, in various circumstances (structural steel in buildings, cylinders in motors, etc.). We also have descriptions of the interrelations of gaseous states with liquid and solid states in various arrangements (e.g. pneumatic systems.)

So quite frequently we can predict how a composite relational system will interact with another composite system of relationships. I have used only a few such examples.

The logical fallacy is in the supposition that the "whole" should have a pattern of interaction with other systems which would be the same as the pattern vis a vis one or more of the parts which make it up.

A factor contributing to conceptual haze is our difficulty in predicting the patterns of activity in highly complex dynamic networks – activity both internal to and external to such networks. Examples of such systems are ecosystems, the global commercial fabric, and, at a lesser scale, the internet.

But we do have some tools of analysis, and they tell us a good deal about such systems. One of my personal favorites, because of the time he spent reviewing with me his work, Robert Ulanowicz (1997), has used information theory, careful measurement of ecosystems, and some deep insights by Odum (1983) and others to illuminate key features of how ecosystems work. A number of investigators busy themselves with internet network phenomena. A few economists are addressing economic systems from complex network perspectives. Eric Schneider, from his highly useful perspectives, also urges this sort of approach.

We can predict the activities of complex network phenomena in many cases. As is the case with human bodies and brains. So we are not entirely at sea in the compositional stage of complex systems, and complex adaptive systems.

Moving from the composition step to the external, emergent effect step, we can see the emergent effects of complex systems on other things interacting with them in many cases. E.g. we can rather directly see that the recently rapid expansion of the activities of human social systems has extinguished some animal and plant species, diminished others, and shrunk some forests. Now it appears that our social system is heating up the earth's atmosphere.

We also tend to confuse ourselves when we make the over-general assertion that we must use different languages to describe interactions at different levels of aggregation.

It is true that we have devised different languages, mathematical and verbal, to describe interactions within hadrons, between hadrons, between molecules, between rocks, between liquid and solid states, between planets, between lovers, etc.

But the power law expressions, denoting, I suggest, correlational activity, operates at all scales from hadron up. The Boltzmann-Gibbs-Shannon-Ulanowicz types of order/entropy measures operate over vast ranges of aggregations and applications.[16]

We also know that gravity works as to all physical bodies, whatever the level of aggregation, that electromagnetic forces, as we now conceive them, are involved at fermion levels and upwards, and are pervasive in the functions of aggregates tiered from atom up.

One of the more interesting questions relates to "top down" effects, or "downward causation" in emergent phenomena. This can lead to considerable confusion, when words such as these are used without careful matching with physical processes.

I suggest that we differentiate between three different ways of ascribing a top down or downward causation effect. They are
1. fundamental mechanics of the universe, such as the Pauli exclusion principle, conservation of energy, gravitational constant, Planck's constant, charge, polarity, etc.

---

16 For example, we find this technique used at the quantum level, by Rovelli, as previously noted, by Shannon and followers in a vast range of communications applications, and by Ulanowicz in defining the degree of order in ecosystems. Boltzmann expected universal application of his formula. I include Gibbs entropy in this reference and posit a genre of entropy measures, to take account of the fact that measures of the total states in a system and the distributions of states in a system must take into account correlations if they are to reflect ordering in systems.

2. the connection between a state and a state or states from which it eventuates and on which it is conditionally dependent. (An example would be the Earth's orbit and composition as the preceding state, and then the development of life in the ocean as eventuating states. In the evolution of life forms we can trace the connection between the structure and function of progenitors and their descendants.) and

3. The actions of living systems having the capacity to direct the energy flows within themselves so as to affect or manipulate the relational set in a chosen field of focus.

The first two categories are sometimes characterized as "boundary conditions".

The third category, I will later suggest, is the one which engenders many of our concepts as to causation and top down effects. In later chapters on life organization and causation, we will touch briefly on this, including how life systems create and react to "mutual information" sets.

We can easily observe that in highly complex living systems with internal decision making systems (input, processing, output), the organism as a whole can intervene in the operations of its constituent parts. We manipulate our own bodies. We can also see that the organism can act on other objects. This is an "emergent" effect of the organization, and it also is a "top down" (actually recursive) effect of the organization upon itself.

The same is the case with complex organizations made up of humans. The "State" – the human-made aggregate system -- can act upon and regularize the functions of its human components. This is "top down", recursive activity.

These actions or causative effects thus can with some clarity be said to be "top down" with respect to elements within the complex system which acts upon itself.

In a different use of the "top down" language, such actions or causative effect might also be said to be "top down emergence" with respect to other systems affected by the system of a life unit or a grouping of the life units. Other (let us for example say inanimate) systems are manipulated by the living system, rather than by themselves.

However, it would appear to be conceptually cleaner to state that a complex aggregate with the ability to employ a variety of relational modes acts upon another relational system (which may or may not have internal organizational capacities) so as to modify the internal relational system of the other.)

As noted above, the top down effects of a living system on its components, and on other things, can be seen as an effect of the self referential aspect of the universe's creation of complex organizations which can act back onto the universe.

That is, humans have long observed that in living systems the Universe observes itself and acts upon itself. (With humans, we extensively use representation systems (language, graphics, computer coding), which are also exercises in self-reference on the part of the Universe.)

There is nothing in such "top down" activity which should be seen as entailing a departure from the basic laws of the universe explored in our disciplines of physics, chemistry, etc. Nor is there any need to invoke any substitution of other, arcane "forces" or factors to explain these phenomena. The blade of the scalpel, or the jaws of a nail clipper, operate by means of extensively characterized physical forces. The bars on a jail cell likewise. I have not seen those investigators whom I most respect suggesting any arcane or magical new physical forces in "top down" activity by living organism on themselves, and systems external to themselves.

Thus, to see top down causation in complex dynamic systems all we have to do is to imagine living creatures as dynamic processes with sub-processes for registering input, and processing such information in ways which channel the activities of the organization so as to produce "emergent" effects – as defined in this discussion -- on themselves and their surroundings.

Then we can add the self referential aspect of the universe which created the process (This, among other things, means that the complex process is itself, in its sensing and in its effects, a part of its surroundings. Confusing perhaps, but this self-referential universe business requires some interesting bookkeeping, as Godel pointed out.).

To recapitulate, it is difficult to see "top down" effects when we focus on the immediate act of formation of simple liquid and solid systems. The components lock into such relational regimes, and are held in place by the act of correlation. They do so, in most circumstances, in which they are brought into close proximity, and lose enough "heat energy" to undergo phase transitions.

We might see a primitive, elemental form of "top down" effect, if we wish to, in the Benard cell situation.[17]

But we can easily see "top down" actions in the interventions upon external systems of complex, dynamic organizations which have within themselves dynamic internal ordering systems.

Let us defer dealing with a number of other types of situation in which "emergence" language has been used in the conceptual framework of complex adaptive systems, to the chapter dealing with life organization on a more extended basis.

---

17 The elements in the liquid which are involved in a Benard cell are correlated to a degree among themselves – they have reduced degrees of freedom with respect to each other. As the liquid in the cell moves in a coordinated fashion, this coordinated or correlated system carries along its components, so as to make them move relative to elements outside the cell.

If we could put a decisionmaking circuit in the Benard cell - sensing or registering surroundings, correlating such info and internal action responses, and organizing internal systems to act on external circumstances -- we could have it act on other cells, and with sufficient complexity act upon itself. But that is several steps down the path of organization.

The simple concepts set out here go the basics generating forms over the evolutionary process. If we recognize that correlations – patterns – in dynamic processes are an expression of energy minimization in such processes, that the Pauli exclusion principle lies at the bottom of shape in the first place, and that shapes express underlying symmetries of the universe when they precipitate out in phase transitions, we can do without some of the explanations elsewhere offered for pattern formation (or "form", or shape formation).

We are powerfully inclined to identify the process of evolution as an emergence machine. Indeed, it is. It is also a realization of a special case of the progressive realization of the potentials of the Universe to create ordered states. I have set that out in the first three chapters, and the earlier part of this chapter. We can keep better track of this process, I submit, if we use the format here set out -- combinatorial, part-whole, stepwise, hierarchical realization of relational regimes.

The system of reproduction evolved on Earth is a marvelous, precious, and, if you will, sacred (to its beneficiaries) embodiment of the universe's creativity. Let us focus on that in the separate chapter on the life process.

5 Causation

The preceding materials provide means of simplifying and organizing concepts of "causation." Causation, in our universe, is a combinatorial event in a linearly unfolding, probabilistic universal process.[18]

The picture of the universe which I have proposed leads to a view of "causation" similar, though not identical, to that of John Stuart Mill (1843), who suggested that cause is the antecedent, or the concurrence of antecedents, on which [a given phenomenon] is invariably and unconditionally consequent. It also has similarities to views attributed to the Stoic school of thought in ancient Greece.

Given advances in science's depiction of the mechanics of the universe since ancient Greece and since Mill, we can refine and add to such concepts, without abandoning them.

As pointed out in the preceding chapter, one can start with the fundamental mechanics, or invariants, of the universe, so far as we today can characterize them, such as the Pauli exclusion principle, conservation of energy, gravitational constant, Planck's constant, charge, polarity, etc.

Then one conceives of a stepwise correlational process, operating on the invariants, which evolves levels of aggregation, and relational interactions between the aggregates evolved – i.e. atoms, suns, galaxies, planets, molecules, etc. This system builds ordered states in a stepwise barrage of realized conditional probabilities. (Thus I would not have put the "unconditionally" word in Mill's approach, or formulation.)

In observing and characterizing this process, one can use familiar formulations of force, mass, gravity, etc. to explain what will happen in a given situation. If galaxies collide, we can map the interactions. If and as suns form, we can predict radiation types, the building of complex atoms, supernova explosions, etc. We can thus do extensive if/then modeling.

The over-all picture is one of process, working off invariant metrics, creating ordered phenomena as it goes. One can use the language of universal machinery (as in the 19th century) or information processing (in early 21st century language). Either way, one comes up with, simply, a process of the eventuation of relational states and interactions.

One can map traditional ideas of causation onto this process, as I did in the prior chapter in referring to the connection between a state and a state or states from which it eventuates and on which it is conditionally dependent. But it appears to me that the process is the picture to keep in the center of attention.

---

18 This discussion proceeds from the assumption of the dimensional structure of our universe as we experience it. For a discussion of how more dimensions or fewer dimensions, including a 'time' factor, would exclude causal relations, see a short, though significant, paper by Tegmark, Max (1997).

This process carries through in the evolution of life forms. But we human life forms have a particular point of view which we tend to impress upon the working of the universal process. We are a manifestation of the evolution of what Stuart Kauffman (2000) has called "autonomous agents" -- complex systems with sensing systems, decision systems, and systems for acting on the environment of the complex system. We are a part of the universe which takes a particular, dynamic, assertive point of view as to the rest of the universe (and as to itself as well).

This (somewhat) autonomous agent may be advantaged by being able to move this way or that way, or to push this or that "button", so to speak, in its context. If the autonomous agent (not entirely autonomous of course – just having some degrees of freedom to exploit) becomes as complex as humans, with language and collective action systems, it (or they) can construct maps of what happens if/when it/they act upon the "environment" – the set of relationships – in which it/they are embedded.

We humans are thus directly led to concepts of "cause" in terms of -- I/we do this -- and "effect" -- we then see this happening, or that. (However, given how often we run into the "law of unintended consequences", we may often see something happening which is different from the "effect" we intended.).

Thus, in the preceding chapter category in preliminarily dealing with concepts of causation, in the context of "emergence" theories, I suggested treating as a separate subcategory the actions of living systems having the capacity to direct the energy flows within themselves so as to affect relational regimes at various levels. This is a special case of the universal process, having forms of manifestation particular to itself.

Since this "autonomous agent" lives in and is subject to the metrics of the universe which created it, it must at least accord in some fashion with those metrics. We spend a fair amount of time trying to understand, to map, so to speak, those metrics. What I suggest here is that it may help a little to build into our understanding that in speaking of "cause and effect" we may project our "autonomous agent" framework onto the universe, and to mix up, or confuse, categories and descriptions of what we do with the broader themes of how the universe itself operates.

The Universe, I would suggest, just grinds onward. The universe does not visibly try to "cause" anything. It just operates according to its rules, in a probabilistic, combinatorial, computational, continual fashion. We, little bits of the universe endowed with a few degrees of freedom to work with, try to "cause" and to get "effects." [19]

Further, our views of cause and effect tend to be derived from our particular vantage point in cosmic evolution, to considerable extent. In our own evolutionary

---

[19] Of course, as we track back through the unfolding of the Universe's ordering system, we seek, in the historical terminology, the "first cause", or set of factors, and have come to what in recent terminology sometimes is termed an initial cosmic event, or singularity, or frequently but not necessarily "big bang". Lee Smolin is a good source of information on this line of inquiry. I do not attempt here to add to his observations and suggestions.

history, we have tended to operate largely in condensed matter (high correlation) situations. This allows us to attempt to assign probabilities to anticipated or projected events. But we also operate in settings with differing degrees of organization, or probabilities, in situations in which we have incomplete information on our settings, and in situations where the possible range of combinations is sufficiently large to pose bookkeeping problems for us.

In addition to acting in and upon complex systems, we have some, much less than complete, perceptions of our mental states and mental categories. We often seek to relate our internal states to, if not to impose them upon, external circumstances. Often, given the complexity of our surroundings, and lack of clarity about our own internal processes, we have not been very clear in our bookkeeping as to internal and external events.

As just one example of unclear bookkeeping, from what could be a very lengthy list, one observer, Hulswit, notes,

> "It was Mill's view that what we usually call the cause of an object or an event is only a partial cause. In ordinary discourse we tend to call the cause the factor to which we wish to call attention, although it is not the only factor. We select one condition out of a whole set of conditions which are together sufficient, and call it the cause. What we call the cause is: (a) the last condition to be fulfilled before the effect takes place, or (b) the condition whose role in the affair is "superficially the most conspicuous" (1874, 238-39). Thus, we say that striking the match caused it to burn, because it was "the one condition which last came into existence" (proximate event). And we refer to the gene as the cause of the cancer, because it is the most conspicuous of all the conditions involved."

We are not always so ham-handed. Allow me to call on the law, an occupation which I pursued for several decades. In legal contests over who in a social system should bear the burden of unfortunate, damaging occurrences, the courts generally recognize that multiple factors could contribute to the occurrence, in the sense of being necessary to its happening. Courts often look to the "proximate" cause (a closely related condition or act, with direct links to what happened), and try to determine whether as a society we could have expected a member of the society to have acted so as to avoid that "proximate" cause (these expectations are "negligence" and "standard of care" considerations).

The law has usually been pretty cogent, in Western venues, within the scope of its objectives. However, the law's objectives are in large part to sort out human squabbles and keep human order, rather than to explicate universal order.

In more comprehensive, academic treatments of causation, we see references to the internal organization of the living entity, or a society of them, such as "goal seeking", feedback loops, and (designed) homeostasis arrangements. This has largely to do with how living creatures work.

In sum, to recapitulate, in dealing with such conceptual arrays, and in some cases tangles, this treatment of the origins and development of ordered states in the universe, as related to our concepts of "causation" in effect updates the Stoic and Mill approaches to notions of causation.

In general thrust, I suggest, the Stoic/Mill tradition, or sustained thread of conceptual organizing, has served us well.

However, this update:

1. parses out some apparent invariants in the universe's function mediating, or serving as "boundary conditions" in the evolution of ordered, computational, compositional states, in which there can be preceding and succeeding states;

2. provides a structure for the evolution of ordered states within a given set of "boundary conditions", in the hierarchical, conditional probability, if/then, antecedent condition/subsequent condition development of ordered states;

3. clarifies the standpoint, or viewpoint, which gives rise to many of our human mental constructs about causation, providing a better basis for organizing those constructs; and

4. Is designed to bring into clearer focus the range of probabilities which may come into play in tracking, and seeking to influence or affect, the sequences of events in which humans participate.

6 The Arrow of Time

In the pre-scientific era, our species invented a variety of idiosyncratic suppositions about the future of earth, humans, and however they imagined what we now would characterize the cosmos. These today warrant only historical reference.

In the immediately preceding five centuries, and particularly in the last two centuries, rapidly developing empirical investigative tools and conceptual tools have yielded a number of different scenarios for the universe. These include a steady state universe, as proposed by Fred Hoyle and others, a "heat death" in which all eventuates in uniform randomness (a popular interpretation of some time ago arising from Boltzmann's work on thermodynamics and entropy), cyclical expand-contract-expand universes, multiple universes (or multiverses) of various versions, cold, dark and isolated uniformity, disruptions of all forms of matter, and more. An overview of such projections impresses upon the observer both the ingenuity of the physicists at any given time, and the continuing flux in conceptions.

One of the conceptual elements of this mix has been the issue of whether there is an "arrow of time". There are many renderings of this issue. A sketchy but illustrative sample of approaches, for general readers, as to the "Arrow of Time" can be seen on Wikipedia (2016)

A point of focus has been the fact that several "laws" of physics appear to be time-symmetric. Another focus of attention is the "second law of thermodynamics", which has appeared to predict inevitable deterioration of ordered states, or non-equilibrium combinations of discrete elements in a state set. The second law, in particular, is a formidable construct, and has thus far defeated a great many attempts to deconstruct or evade it.

The construct of order generation which I offer here embodies a clear and direct arrow, associated with our concepts of time. This picture points our attention away from the reversible interactions of particular particles, and toward an evolution of the over-all state of the Universe in which subsequent states of the universe differ from prior states on a progressive and thus far non-reversible (or at least not universally reversed) basis.

The conventional explanations of "Big Bang" and related cosmologies present a picture of an expansion of and related "cooling" of the universe. This entails a cascade of phase transitions involving fundamental forces, quark-gluon transitions to simple atoms, complex atoms, simple and complex molecules, and aggregations of these entities into galaxies, stars, planets, etc.

Strictly speaking, phase transitions can both manifest correlation and differentiation events, in the direction of diminution of degrees of freedom, and on the other hand in the opposite direction manifest disruption of correlations and differentiation.

However, the expanding universe picture presents a framework in which there appears to be a direction of phase transitions toward differentiation and correlation, in the universe as a whole, over "time" -- that is, the continuing evolution of the Universe.[20]

As I have noted earlier, David Layzer and Eric Chaisson have provided slightly varying but compatible explanations of how the expansion of the universe allows ordered, or correlated, relational regimes to arise and persist, notwithstanding the second law of thermodynamics.

Layzer (1990) spoke in terms of the rate of expansion outrunning the rate of equilibration involved at local scales, while Chaisson (2001) summarizes the argument as "In an expanding universe actual entropy ... increases less than the maximum possible entropy"-(p. 130) -- thus allowing for, or requiring, ordered (negentropic) relationships to arise and persist.[21]

Both explanations of order generation posit a process of "expansion" and generates the phenomenon we call order (as that word is used in this monograph -- the visible, sensible universe), on a one-way, continuing basis.

Chaisson depicts the universe as a non-equilibrium process in which energy flows into and through ordered systems, such as galaxies, stars, and life processes. This provides a cosmological basis for non equilibrium thermodynamics, in terms which unite non-equilibrium thermodynamics language and relational analyses. Patterns of processes arise and are evident as ordered, dynamic relational regimes.

Thus far, all of this is based on extensive observations of the universe, and currently respectable theories offered by credentialed academics. Order, or organization, obviously has arisen in the universe, some of it persists, and this compels some attempt to account for it.

The account, or explanation, which I offer here has a number of aspects which might be thought to be challenging.

First, to depict the tangible, visible universe as the manifestation of progressive, hierarchical correlation and differentiation processes is equivalent to depicting the visible universe as both marker and embodiment of an "arrow of time" (again, a process which we benchmark by measures of process which we call time).[22]

---

20 At least at the level of aggregation of baryons, each correlated and differentiated system thus evolved can be considered, from the relational point of view, as a network of relationships, as noted before.

21 Though I cannot fault Layzer and Chaisson as to how the books between "maximum" entropy, "actual" entropy and "negentropy" must be balanced, the reader of this narrative might infer I am led to consider an active, or shaping, role in the historical evolution of the universe for the "negentropic" process of ongoing correlation processes.

22 I note here that in his "Investigations" book, (Kauffman, Stuart (2000), on pp 249-252, Kauffman articulates a beginning of an idea as to how "decoherent" phenomena preferentially build relationships with each other, and thus co-construct a universe. That is in substance what I am proposing in this manuscript. If one adds together Kauffman's intuition, Rovelli's work, and Barabasi's tracing out of "preferential attachment" models, one approaches the themes of this manuscript.

Secondly, in parallel with the first proposition, the arrow of time is marked and in a sense motivated by the expansion of the universe, and can be characterized as a not-visibly-reversible thermodynamic process at the scale of the universe.

Thirdly, if the foregoing is true, the passage of "time" -- which can in this construct be considered as the unfolding process of the universe -- would create an accumulation of ordered, or differentiated states in the universe.[23] This would appear to be evident, in gross observation, as the universe has created complex atoms, and tiers of combinations of those atoms which I have characterized as tiers of correlations of correlations.

A naïve characterization, or formalization, of this process might be something like $O=A*Exp*T$, where O equals accumulated order, A is a presumed constant, Exp is the expansion rate, and T is elapsed time, or process, in this universe.

This third perspective, though based on extensive observation and currently plausible theory, raises numerous questions, or, at least, issues.

Does the amount of ordered states in the universe in fact accumulate indefinitely, or is it, perhaps by operation of black holes, limited, or an equilibrium quantity?

Will "the universe", as many suggest, become relationship islands, where event or relational effect horizons limit the scope of the relevant universe -- the area in which interactions can be perceived to be occurring -- as to each locus of relational realization? (However, in this sort of setting, there might arguably be perceived to be effects of the Universe in gross which might be explained only as effects of a virtual infinity of other relational islands.)

Can the variations in the rate of expansion of the universe be correlated with discernible variations in the creation of ordered states?

Lastly, as to what this evolving universal process is producing, over "time", in terms of our beloved complexity, the manifestations of energy dense complexity would appear necessarily to be a very small proportion of the extant universe, for any foreseeable future. The construct which I here offer does not seem to provide a basis for projecting a substantial and ever-growing quantity of complexity overwhelming all other aspects of the Universe, which, as humans, we would like to see and participate in.

This limited role for complexity is necessitated by order being realized conditional probabilities, following power law statistics – with the necessary dynamic that each next step of complex correlation can only be a fractional realization of the volume of the preceding stages.

But then, in consolation, we appear to be in a very big Universe, as compared with us, so that a miniscule fraction of complexity could be a very large total of complex

---

23 As I read his "Time Reborn" book, Smolin (2013) also treats the universal process as an historical, existential entity, drawing a distinction between it and the depictions of 'spacetime" which present a pattern in which the unique existence of the process is submerged. But Smolin does a great deal more with how time is treated in physics than I attempt in this brief overview of the "arrow of time" issue. Indeed, Smolin comes up with a plethora of arrows.

states, as compared with us on Earth today. If we could locate ourselves, quantitatively, on the complexity curve later offered, and reliably assay a total mass of the Universe, we might then be able to estimate the totality here referred to.

Lastly, in this Chapter, is a note on what is considered "time" for the purposes of identifying an "arrow" relating to the unfolding of the visible universe. In this treatment, the precept, or concept, that the "laws of the universe are the same everywhere and at all times" (here taken as instances of local, relational observations or interactions) is observed.

Although this may seem to be a resurrection of Newton's everywhere, universal and absolute "time" concept, it does not invalidate the calculations relating to how widely separated local events (or alternatively local inertial frames, or co-moving inertial frames) can have relational effects on each other, and calculations as to how the rate of deviation from "comoving" relationships will affect the registration of events in another "inertial frame", or relational frame, and the one of our inhabitance.

It would seem necessary to maintain the assumption or proposition that there are universe-wide processes of identical sorts and rates in the general evolution of the universe.

7 Complexity

The idea of complexity has generated a large number of definitions and approaches. In recent decades, as areas of sophisticated academic and commercial activity have grown, manifestations of complexity have engaged close attention in many settings.

Exploding computer and communications use have led to numerous attempts to match up computational techniques and complexity in various guises and conceptions. Claude Shannon was an early and successful pioneer in this area.

Attempts to deal with complex economic and organizational systems spurred an attempt to construct a theory of systems which had sufficient breadth and penetration to create general theories of systems applicable across many fields of endeavor. An early organizational lead in this attempt was the International Society for the Systems Sciences, which is still extant.

In recent decades, many scholars and others have focused on "complex adaptive systems" -- assemblages of living creatures, or "adaptive agents", making up systems which themselves can match up with their environment and reorganize internal operations to system advantage (think any of several levels and types of human systems – companies, states, nations, etc.)

Ideas of network systems and non-equilibrium thermodynamics have led to attempts to re-cast traditional forms of economics.

Leading research entities, including but by no means limited to M.I.T. and the Santa Fe Institute, are seedbeds for a host of disparate and often idiosyncratic assaults on various manifestations of complex systems and behaviors.[24]

As in previous chapters, I do not attempt to chase all the rabbits one sees in the field. Given all the heterogeneity out there, I attempt to start with the fundamental organizational themes of the universe -- as is evident in prior chapters -- and to show how what we perceive as complexity develops from and can be related to those themes. If and to the extent this is successful, we will eventually have a common basis for defining complexity, measuring it, and dealing with it in various circumstances.

I submit that the depiction of the Universe's operation suggested here provides a clear and direct starting point for a concept of complexity, though the best means of formalizing this approach, and reconciling the variety of current approaches, will be a work in progress.

That approach is to consider complexity as the number of distinguishable relational states in a given system per unit of time, per unit of mass. This measure will, I submit, necessarily be proportional to the "energy rate density" of the system.

---

24 I have picked out what I consider to be a few of the most important areas to comment upon. If the reader wants to roam, one can go to Wikipedia's disambiguation page on complexity -- they have a charter to try to take into account whatever appears on the landscape --and follow the links. That will lead you far away from this distillation.

This is the condensed formula. It is writ very large in its eventuation, of course.

We start, as in prior chapters, with depicting the Universe evolving, or elaborating. generally, and locally, an expanded, hierarchical body of relational regimes. The elaboration of those relational regimes is the development of complexity.

The logic is straightforward, and, I submit, inescapable. A core element in perceptions of complexity is, to put it simply, a lot of different relationships. In the earlier chapter, I pointed out that the aggregation of nucleons in atoms developed the vast potentials for distinguishable combinations of 92 different atoms. The groupings of these combinations into further hierarchical combinations created elaborated relational structures.

This is the austere, spare, conceptual way of stating the process.

Chaisson, fortunately, provides us an overview of the vast, compelling, tangible drama of the process. In his book "Cosmic Evolution" Chaisson does not explicitly use the correlation mechanism as a key motivator in his scheme, in the way I do here. Nonetheless, he has worked out a sweeping and compelling outline of how this process has developed systems, or manifestations, which we universally recognize as complex, including particularly life systems. In my view, Chaisson's system is closely consistent with the picture I propose, and his insights have been very helpful in developing it.

Let us take a few physical examples of these elaborated relational regimes, before further returning to matters of theory, or overview.

In some worked metals, we can see with the help of a microscope multitudinous "grains" -- patches of regularities of atom combinations -- of varying sizes and orientations. We find it natural to characterize such structures as complex. On the other hand, a perfect, or near perfect, diamond, with its very regular and tightly bonded arrays of carbon atoms, is seen as much less complex.

In dealing with solids like this, we are dealing with frozen computations, so to speak. Let us go to dynamic structures for more elaboration of relational states.

There are simple dynamic relational systems, like galaxies, bathtub water swirls, the Red Spot on Jupiter, hurricane swirls, etc.

Living systems present the most complex dynamic systems we are privileged to observe. In them we see an enormous range of differentiation and gradation of relational regimes – fluids like blood, cerebrospinal fluids, lymph; solids like bones; substances like cartilage with varying degrees of flexibility and rigidity; gas-fluid interchanges; long nerve fibers; muscle fiber strands interwoven; the DNA instruction set at the core of cells; etc.[25]

---

25 One may note that complex life systems on Earth evolved first in liquid phase settings -- a leading candidate site is near volcanic vents in the oceans -- where relational structures could be developed and altered with some facility, in the presence of high energy throughput -- as distinguished from atmospheric gas or tightly bound, immobile solids. From there, life gradually exploited combinations of gases, solids, and various levels and types of constrained liquids or semi-solids. We might expect

Thus, the greater portion of various areas of scholarship as to complexity have to do with life units, and organizations of life units.

Of central analytic importance, in terms of the basic architecture of these dynamic complex relational systems, we see degrees of constraint and degrees of freedom intricately interwoven, and phase transitions in orderly, repetitive cycles, throughout the systems, as Smolin noted.

And of course between simple dynamic systems and the extremely intricate relational structures of life units, we see networks of intermediate complexity, such as economic systems.[26]

I have proceeded from a physical ground-up approach thus far. Others have seen the same elephant from other angles. As communities of investigators and analysts have had increasing encounters with manifestations of complexity in the last hundred to two hundred years, many of the most perceptive analysts have in effect pointed toward anchoring complexity concepts in the relational approach which I here advocate.

For example, a prominent leader in "systems theory" development, B. H. Banathy (1996), defined this perspective for the Primer Group of the International Society for Systems Sciences,

"The systems view is a world-view that is based on the discipline of SYSTEM INQUIRY. Central to systems inquiry is the concept of SYSTEM. In the most general sense, system means a configuration of parts connected and joined together by a web of relationships. The Primer group defines system as a family of relationships among the members acting as a whole. Bertalanffy defined system as "elements in standing relationship."

Similarly, Robert Ulanowicz (1997) has, as I have suggested before, analyzed ecosystems on Earth, universally recognized as complex assemblages, as relational systems, applying the tools of information theory, in "Ecology, the Ascendent Perspective".

Lazlo Barabasi (2002) has studied a variety of complex systems, including the internet, using relational network analyses, finding power laws pervasively manifested. Among the areas studied are living organisms at all scales, from components of cells, to cells, to complex multicellular organisms. Barabasi has directly linked these phenomena to the mechanics of phase transitions and network development, at all levels. (Or, in the language I have used, the correlational process at all levels.)

An expanding body of "social network" scholarship pervasively applies network analyses throughout human social organization.

---

similar development in other venues in the Universe. This accords with the Chaisson approach to defining high complexity realizations.

26 I put such systems as intermediate in complexity, if we look to the structure of the system of economic units, and do not take into account the complexity of the life units making it up.

Thus the combination of these various strains of thought in the basic formulation that complexity can be characterized as the number of differentiated relational states in a system under study, or in the field of focus, per unit of mass, per unit of time.

We do need to define the system. As Lee Smolin (1999) and numerous others have recognized, a "system" is itself a set of relational regimes differentiated from other sets of relationships.

In static settings, one could merely attempt to discern and enumerate the distinguishable relational systems. A closely related measure would be the enumeration of phase transitions recorded in the enumerated states. However, static situations would not measure all elements of process.

It would seem necessary to distinguish systems on the basis of mass. Otherwise complexity would scale with simple increase in total volume of any given aggregation.

In dynamic systems, one needs to take account of the volume of distinguishable relational states, and phase transitions, over time (a measure of the total process mechanics).

As noted before, because energy is involved in phase transitions in dynamic systems, measures of complexity correlate with what Chaisson (2001) has characterized as the "free energy rate density" which is required to drive the phase changes. Chaisson has demonstrated at length (pp 151-207) the utility of this measure over a very wide range of phenomena, including suns, planets, life units (little ones like spiders and large ones like humans), cities, fighter planes, computers and more.

In this section of his work, Chaisson makes the major conceptual contribution, in my view, of linking complexity, free energy density, and embodied information. The preceding analysis in this section accords with Chaisson's. I will refer to these concepts later in discussing the questions of total complexity in life, and whether the course of evolution complexity has increased.

One sees in the life process a beautiful realization of the combination of hierarchy and complexity realization. This process will be further developed in the following chapter.

Let us focus primarily on the life process, in outline, without losing sight of the fact that "chemical" precursors of life had to be combinations of atoms, brought to a differentiating Earth with areas of high energy flows (as at ocean vents), in liquid settings allowing for intricate combinations and recombinations of elements.

Leo Buss (1987), and many others, have pointed out that there had to be some some relatively intricate structures to be combined in replicating entities before single celled life could appear. Morowitz and Smith (2006) have outlined chemical cycles -- e.g. the "citric" cycle -- operative at the core of life, with a combination of elements and processes relatively few in today's organisms, but complex relative to the general setting of ocean elements.

At the single cell life level, we can see an intricate, active set of processes enclosed by a relatively (!) simple membrane, which maintained the active interior processes, by maintaining the high concentration of elements and probabilities needed to sustain function and reproduction. In terms of "emergence" thinking, the cell membrane serves to protect and maintain active, repetitive patterns inside the cell, while presenting a "whole" to whatever is outside the cell. The cellular whole would have a much more random set of encounters in the early (and present) ocean. (This is not to say that there would not be some complexity in cell/environment interactions, but less than in the cell interior.)

As will be noted in a following chapter, group-making -- hierarchical -- processes led to combinations of cellular life units in more than one fashion. But one of the lines of development led to multicellular creatures, like plants, animals and fungi, having self-enclosed reproduction systems. As the sponge illustrates, early versions of multicellular life combined active internal cell activity with relatively simple over all form, and relatively simple interactions with the environment external to the structures constituting the boundary of the creature. Over evolutionary time, the internal organization of some multicellular life has come to involve a great deal of differentiation, with active and complex internal systems.

And then some multicellular creatures began to develop groups (e.g. social animals) with active and differentiated energy processing systems -- that is, current human societies,

We see tiers of complexification. Such a tiering of systems can have different levels of complexity at different levels. When we attempt a Chaisson - emulating assessment of energy rate density to upper levels of such tiering, we have a summing, of sorts, of energy rate densities of the components of the tiered system.

More about this in the following chapter.

Let us sum up. There are a number of advantages in grounding complexity concepts in the correlational, hierarchical elaboration of relational regimes described in this series of chapters:

1. Complexity concepts (including but not limited to "systems" theory and network analyses) can have a common physical basis, a common operational rationale, and eventually, I suggest, common mathematical languages. In particular, ideas of computational complexity may be aligned with concepts of the informational nature of the universe, and with Boltzmann/Gibbs/Shannon/Ulanowicz forms of statistical analysis[27];

---

27 I am led to venture into territory much trodden by mathematicians, an hazardous exercise; but will venture an observation about computational complexity, p and np problems, and the like. Stuart Kauffman illustrated a conceptual problem in his book "Investigations" when he talked about the vast field of combinations expressible in combinatorial notation when one dealt with chemical species and their elements, versus the "adjacent possible" realization of ordered states. Kauffman addressed this divide by somewhat different means than I use, but he clearly saw the need to address it, and gave us a pregnant phrase to use.

I have repeatedly contrasted the picture of an entirely random assemblage of hadrons, etc on the one hand and on the other the stepwise hierarchical realization of ordered states which make up the sensible universe. One can venture that the correlational process is the Universe's computational

2. Manifestations of complexity can be put in an hierarchical context or framework, clarifying their phenomenological statuses, and their statistical abundances;
3. Concepts of complexity, information density, and free energy density are closely linked.
4. The conceptual problem of seeing random arrangements as highly complex simply in their randomness (because they can assume so many configurations) is solved. The relational system approach clearly identifies the necessity of constraint (correlation) in creating relational systems. (One can still include in complexity concepts or measures random elements in constrained contexts – e. g. gas in lungs, in pneumatic systems, etc.)

The foregoing advantages lead to seeing the manifestations of complexity of greatest interest to humans (e.g. complex adaptive systems) as
1. Statistically very rare developments in the power law distribution of complexity realizations in the universe,
2. Operational in the processes conceived of as "non-equilibrium thermodynamics" (i.e. ordered states arising in processes, a zone in which life is manifested); and, not surprisingly,
3. Realized in the rare but to us endlessly fascinating manifestation of the life process, now to be addressed.

---

economizer, in that it fixes the realization of combinatorial "possibilities" into constrained, consolidated state spaces, and limits further combinations to the interaction of those constrained states with other states, up the hierarchical structure of the realized universe. This occurs progressively from the quantum level up. Put another way, correlation fixes probabilistically adjacent possibilities into operational realities.

I would therefore suggest looking for and constructing "depth of computation" models and other mathematical models of the universe which follow this template and this logic. At each level of hierarchical organization, the computational realizations of the components is subsumed, as well as their computational operation in creating the composite, and the next level of "complexity", and computation, is brought into being in the relational system of the composite with other distinguishable entities.

8 Life

At last we arrive at that which most interests us – ourselves. Let us approach the examination of this beloved subject by assessing first fundamental characteristics of the life process, within the framework set out up to this point.

Drawing on the characteristics of the ordering process which created life, I will offer a few propositions about the characteristics and directions of the life process on Earth. First I will assay a capsule definition of what fundamental characteristics distinguishes living things from inanimate things. Then I will propose that evolving life on Earth:

1. exploits combinatorial potentials;
2. exploits both structure and randomness;
3. works off the correlation dynamic and energy flows in sustaining and generating biological forms over time;
4. builds groups and hierarchies of groups;
5. creates and exploits "mutual information" between reproducing units and the life structures created by the reproducing units; and
6. builds a set of structures – coexisting relational regimes -- tending to realize the possibilities of the life process.

The signatures of the ordering process which I have before described are evident throughout the development of the life process. Also evident in the panorama I paint are some peculiarities of the process of building relational regimes on reproducing elements in dynamic energy flows.

This picture will have elements long recognized by scholars of biology. For that reason, in some cases I will only touch on well recognized themes, to put them in the context of the over all scheme offered here.

The picture does differ significantly from some recent paradigms offered in this field. Thus I will suggest that the thermodynamic order building view of evolution offers a necessary complement to and context for the gene-centered view of evolution, particularly "selfish gene" concepts, and requires reorientation of some of the approaches previously taken in trying to identify and characterize complexity in the arena of life.

In filling out this picture, I propose a definition of humanity's place on life's organization chart. And this discussion will lay the premises for discussing humanity's potential roles in the further development – or perhaps temporary degradation – of the life process on Earth.

8.1 What is Life?

We have seen millennia of commentary on the nature of life. There are today a wide variety of probes as to how life got started, and some of its salient characteristics.

And of course there is a formidably large literature on the general subject of biology.

In this chapter I put life in the setting of the universe's evolution of complex ordered relational systems, as a manifestation of "non-equilibrium thermodynamics".

In this context, a schematic definition of life is, somewhat surprisingly, not all that difficult. A number of observers (including but by no means limited to myself) have in recent decades converged on thoughts which can be put in compact, schematic language.

In barest outline, given the background of the preceding chapters, one can say that life units intake "energy" into patterned dynamic processes and reproduce. We can proceed from there.

Brooks and Wylie (1984) probed this territory, in somewhat complex language, in their book "Evolution and Entropy" (note particularly p. 102). Shortly thereafter, John Collier (1986) wrote an article, entitled "Entropy in Evolution", to defend Brooks and Wiley from critics of some of the major themes in their book. In doing so he adopted the Brooks and Wiley approach to life definition.

In 1999, Lee Smolin included a chapter entitled "What is Life" in his book "The Life of the Cosmos". After preliminary discussion which establishes the necessity of energy flows and non-equilibrium thermodynamics, Smolin advances, on page 157, the formulation that life units are self organizing non- equilibrium systems[28] which have their processes governed by a program which is stored symbolically, and both the program and the process reproduce.[29] Smolin filled out the picture in more detail, reflected in this footnote.[30]

In his book "Investigations", Kauffman (2000) identified life as entailing "autonomous agents", and characterized such agents as embodying "an autocatalytic[31] system able to reproduce and able to perform one or more work cycles" (p. 47.)[32]

---

28 The preceding discussion makes clear that Smolin is using "non-equilibrium" to refer to a product of "non-equilibrium thermodynamics". As I have elsewhere noted, dynamic processes can themselves have equilibria, and life units demand such.

29 I doubt the program element is a necessity as to all forms of life in the Universe, but on Earth it has been a great convenience.

30 "A distinguishable collection of matter, with recognizable boundaries, which has a flow of energy, and possibly matter, passing through it, while maintaining, for time scales long compared with the dynamical time scales of its internal processes, a state-configuration far from thermodynamic equilibrium. This configuration is maintained by the action of cycles involving the transport of matter and energy within the system and between the system and its exterior. Further, the system is stabilized against small perturbations by the existence of feedback loops which regulate the rates of flow of the cycles." Smolin (1999) p. 135.

31 By "autocatalytic" Kaufman (and Bob Ulanowicz, who also developed this concept) means a sequence of chemical interactions which loops back on itself. E.g. A->B->C->A.

32 I generalize Kaufman's "work cycle" requirement into the requirement that the system be topologically circular -- a condition necessary to stability of the system (as Kaufman in effect recognized in his discussion of his "work cycles") and fulfilled throughout life systems. That is, a cell, and any

I came up with the succinct formulation proposed here back in the late 1990's, and exposed it to public view in a presentation to the Institute for Liberal Studies in 1999. At that time I was unaware of the Smolin, Brooks and Wylie, and Collier works. Nor did I know that Kaufmann had a book in process which would approach the topic from his "autonomous agent" perspective.[33] I thought I had put myself out on a limb, unaware that others had also been building and buttressing the limb.

Looking back, several persons were seeing the same elephant, from slightly different angles. That sort of thing happens pretty regularly.

This schematic language, subsuming but not detailing the chemistries of metabolism and modalities of reproduction, has schematic sorts of utility.

Let us take the energy requirement. As Lee Smolin in effect points out, the energy levels need to be, as to biological life units, scaled to those which assemblages of carbon based molecules can absorb and process. Today, the bulk of the life's energy gathering comes via photosynthesis – converting the excitation of electromagnetic radiation into the life process. But life units also work off other sources of "energy". Microbes and other creatures at deep sea floor vents, where magma oozes up through the ocean plates, use the heat differentials there to power their metabolic machinery. Microbes living in rocks appear to work off local energy differentials resultant from "chemical" differentiation.

And then, of course, life units evolved the trick of consuming other life units, working off the encoded, stored energy in the consumed units. We humans do that.

Thus, wherever we go looking for life, we are going to need to examine energy flows, their levels, and their stability over time. It follows that the characteristics of those energy flows will determine the quantity, degrees of elaboration, and durability of life phenomena wherever found.

If we succeed in building "artificial life", we will have to match the energy input flows to the processing and reproduction system built into that life. We might use electromagnetic waves, direct -- as in wired -- influx of electric charge, mechanical

---

living system, has to have cycling systems which maintain it in its basic organizational format. One can also see the topological circularity in the reproduction system, where we all start off as embryos and go through a (sometimes bizarrely complex) development and conception cycle coming back to an embryo.

33 Viruses pose a challenge to this definition. Biological viruses do the one central act of reproduction. They harness energy and patterned processes --- available in their surroundings -- in the act of reproduction. But we do not observe, in the form or frozen pattern which we call a virus, energy flows in patterned arrangement. With viruses, all three elements of the definition are present, but only when compressed into the reproductive act itself. One way of thinking about viruses is to suggest that they piggyback on what we can see as living entities, using the definition I have offered, by merging into the life definition when they are in a host living system. In other words, they merge into the energy driven patterned process system – that of their host -- in the act of reproduction, and depart from it in frozen form until they can merge again. Life, though rare, marvelous, and some would say sacred, is also a tricky bastard.

energy, ground up human beings – whatever we use, the energy flows will be a determinant of the possibilities of the life forms.

And, as I will later suggest in more detail, if we are to succeed in expanding the potentials of life on earth, and derivatively our own potentials, we will need to increase the energy acquisition of Earthlife.

Our social organizations also and analogously require energy inflows for their metabolism, even though we are not yet to the stage of reliably reproducing societies from societies, and thus our societies far fall short of reaching full "unit of selection" life organism status.

Human societies in the agricultural and early industrial stages drained the global energy reception of the "legacy" ecosystem as a whole. Vitousek and Ehrlich (1986) estimated a few decades ago that we have converted about 30% of the net energy absorption of land on the Earth into the structures supporting our civilization. Then we added the stored energy of prior life forms (fossil energy). Now we seek energy from nuclear fission and fusion, wind, geothermal heat differentials, artificial photosynthesis, etc. More about that later.

Some, including but not limited to James Kay, Eric Schneider, and Dorion Sagan, emphasize the role of gradient reduction in providing energy for living (and non-living) systems. On Earth, life is depicted as working off the thermodynamic differential between the radiation impinging upon it and the cold of surrounding space.[34] Such differentials are produced by the correlation process, which also organizes the systems feeding off them.

Their work, and the conceptual definition of life offered above, is consistent with, and given more ample conceptualization, by work by Morowitz and Smith (2006), who, in a paper done in association with the Santa Fe Institute, suggested that life arose as a means of "...creat(ing) (energy) transport channels in the chemical domain, employing the more concentrated energy flows associated with molecular re-arrangements..." These authors then brilliantly set out a core set of chemical pathways involved in what we call metabolism, ubiquitously present in life processes.

Morowitz and Smith set up a more expansive, but still operational, frame for considering life than I had done, in that they depict life as a thermodynamically compelled (in rare circumstances) means of channeling energy flows between differing potentials, in a way allowing continuance and evolution. This view has implications for our current issues and challenges, which I will outline in following chapters.

The regular, patterned processes of life units give them their shapes and their metabolism – the translation of energy differentials into the pattern constituting the living thing. In complex life, action, or energy, pathways sustain differentiated

---

34 Schneider and Sagan (2005) also maintain an active website dealing with topics relevant to the central messages of the book. www.intothecool.com. (Schneider's views will be cited more than once herein, as he gets more than a few things right, in my opinion.).

structures, or systems, within the life unit. And of course the system has to accumulate the energy and elements to reproduce.

On Earth, these patterned processes were worked out in liquid relational states, rather than either gaseous (no order) or static (frozen order) relational systems. In the liquid system, a semi-solid container system, the cell, evolved and acted as a sort of probability amplifier, in capturing and bringing into close proximity the atomic/molecular/and above aggregates which self-organize and catalyze (impelled by thermodynamic forces) the interaction patterns of life.

Thus, if we want to find or to create life we need to imagine where and how materials and processes can be sustainably "self" organized – perhaps using the same processes as on Earth, perhaps not.

Earthlife's core metabolic pathways have been expanded over evolutionary process, as Morowitz and Smith set out. Those were first worked out in, and given the composition of, the oceans on Earth. But, as noted before, according to Barabasi those pathways generated high uniformity in the relational systems employed, on land, sea and in air, and complete uniformity in the power law distribution of activity levels of nodes and of links.

So we can count on finding similarities of the relational systems in any form of life we encounter, and in any form of life we successfully create or modify.

Replication is the process which sustains and propagates the order possible in dynamic relational processes. Replication creates the evolutionary process, bringing into being the proliferating threads of form and process on the orbiting wet rock we call Earth, in its own constrained trajectory through order-created space, in the process we call time.

Because life would not proliferate and persist without replication, and replication is the vehicle for the unfolding development of life through time, biologists speak of life and evolution as conjoined, as mutually necessary. So now we turn to themes of evolution.

8.2 The Advantages of Combination, Recombination, and Sex

In Chapter 3, I listed combination of elements (into relational systems) as one of the fundamental themes of the construction of order in the Universe. Evolutionary replication has reflected, orchestrated and exploited combinatorics as a part of generating variety and scope in filling out life's potentials.

The exploitation of combinatoric possibilities is, I suggest, very largely what sex is about.

Before organized sex, microbes simply divided, splitting their genetic inheritance 1x1, or pretty closely, between them. But then some swapping of genes as between microbes occurred, in a stand-off sort of process called conjugation. Some microbes worked out sexual reproduction, of sorts. In Eukaryotic cells, a complex mixing and matching of genes evolved.

Currently, in lifeforms using sexual reproduction – count in just about everyone you can see – reproduction involves, at the cellular level, two stages of cell division, which involve reshuffling the genes in a process called recombination. All the cellular reproduction work has to be done very precisely, at scales visible only by microscope, and repetitively. You will note that even at the cellular level, this is an expensive process, and a slow one compared with simple cell division. This implies a high utility in the combinatorial process.

Then the specialized reproduction cells – each having, generally, 23 instead of 46 chromosomes, have to somehow find in the world and join with a matching cell from (usually) some other organism. This entails enormous expenses at the organism level. Animals invest very substantial efforts in the mating process – complex and expensive displays, courting dances, bowers for bowerbirds, diamonds for human females, and on and on. (Just look at the TV ads.) Plants spray the air with pollen, develop lures to attract animals to act as go-betweens between gametes, and so on.

Biologists have probed why, given all this enormous expense, sexual reproduction is so widespread at the multicellular level of life, from several perspectives. An extraordinary biologist, Leo Buss (1987) compiled major suggestions, summarized in the footnote.[35]

I propose a broader explanation inclusive of those Buss has itemized. In the background is the foundation of the combinatoric mechanism in producing ordered states in the Universe. In the biological expression of order building in the universe, we can easily see that the sexual reproduction mechanism steadily cranks out a barrage of variations of combinations into the ecosystem, and, if you will, into the future. These variations do the things Buss catalogues, and continually probe the possibilities of life. This process facilitates the evolution of new life forms, new niches, new ways of going about the business of life. The variations made possible by sex probe life's possibility space, and help construct its reality space. In the successful pursuit of life, combinatorics must be served.[36] More about this later, when we get to the building of ecosystems.

---

35 "Sex is said to be ubiquitous because sex acts as a mechanism for DNA repair...Sex is ubiquitous, according to the arguments ... presented, because it allowed the evolution of cellular differentiation. Sex is ubiquitous, according to Weissman ...Williams... and Maynard Smith because it provides the individual with crucial genetic variation in the face of changing environments. Sex is ubiquitous because it offers to the group the new and advantageous mutations arising in just one if its members. Sex is ubiquitous because sex promotes rates of speciation and decreases rates of extinction (Stanley).... Gould, however, is certainly correct. Sex is ubiquitous for all these reasons. Sex is ubiquitous because it can be favored at the genic, the cellular, the individual, the population and the species level." p. 181.) In addition, one can add the proposal that having two copies of gene strings allows one copy covering for, so to speak, a non-functioning error in another copy – a sort of proofreading advantage.

36 Another major puzzle has been why multicellular organisms are programmed for death. Without going through a catalogue of proposals such as Buss's as to sexual reproduction, a possible answer is that resources given to combinatorial replication in organisms outcompeted resources given to life unit maintenance in unrecombined organisms.

8.3 Life Exploits Both Structure and Randomness

I have suggested that the combinatorial, correlational process of order building is carried out against a background of randomness. One of the more interesting things about life is that it combines and exploits both correlation – call it for the moment structure – and randomness.

We are all familiar with the role of "mutations" in genetic code. Many are neutral in effect on the viability of an organism, some highly disadvantageous, only a few helpful. There is, somewhat weirdly, another type of "mutation" – a chunk of chromosome will occasionally get disconnected, do a 180 degree turn, and get reinserted. We also take into account the degree of randomness involved in the recombination of gene strings in the meiosis stage of sexual reproduction, and in the matching up of gene strings of separate individuals.

All of these things create new combinations. The "selection" side of reproduction sieves out the combinations which do not work out well for the next reproduction in the cycle.

The genetic code system, combined with the constraints of thermodynamics, conserves organization with amazing fidelity in life's enormous quantity of reproductive iterations. But randomness is the background for all ordered phenomena, and the evolution of the life process has also had built into it, and has capitalized upon, a quantity of randomness.

8.4 The Evolution Process Works off the Correlation Dynamic and Energy Flows in Preserving and Proliferating Forms Through Time

I have traced the determinants of form, or structure, to the constants of the universe and the process of correlation, in creating differentiated relational regimes.

The continuity and proliferation of biological forms has been grasped in general outline since at least the time of Darwin and Wallace. An understanding of "non-equilibrium thermodynamics" has been developing over recent decades. Recently, two active investigators, Ursula Goodenough and Terry Deacon, have taken a lead in weaving together concepts of "emergence", thermodynamics, and biological evolution. I am particularly fond of a section of one of their publications which echoes some of the essentials I have tried to point out in the preceding definition of life section.[37]

---

37 "In the language of thermodynamics and morphodynamics, autocatalytic cycles are far from equilibrium dissipative systems that exhibit coherent behavior by virtue of their dynamic regularity. They are usually transient in the non-living world because all interacting systems, are dependent on initial and boundary conditions and proper energy/substrate flow, and these conditions are usually ephemeral. As we will see, life basically works by maintaining the conditions wherein such cycles can operate reliably." Goodenough, Deacon (2008).

In my opinion, Goodenough and Deacon are, among other things, performing a considerable service in getting us past an overly gene-centered period in thinking in biology. Focusing on gene systems has had enormous payoffs to date, and will continue to yield valuable insights in a broad range of fields. In the genomes of life, we see the life process compactly codified. That codification is given operational effect in the way organisms can be seen to operate all around us, every day. And we can trace the history of the codification over geologic time on Earth. Not a bad tool to work with, if you want to understand life.

However, the focus on the gene – or, to get away from a stereotype, the DNA and associated coding systems – has led to concepts that genes are the only replicators (and organisms are just carriers for the genes). This perspective seems to be inadequate in explaining the formation of effective groups of organisms (like social animals, like ourselves), and embodies a somewhat limited concept of the possibilities of the life process.

In the view of Goodenough and Deacon (2008)

> "A coding mechanism is inherently just that – a mechanism, consisting of a set of markers, like an alphabet, coupled with a process that can interpret it. Its interpretation acquires significance only to the extent it codes for an entity – e.g. an autocell feature or an idea – in a way that promotes preservation of both of that entity and of the code responsible for specifying it. These are the features of a semiotic system. When semiotic systems are copied, the capacity to generate more such entities is introduced; when semiotic systems change (mutate), the capacity to generate novel entities arises."

One can briefly illustrate the interactions between life's coding system and the structures for which it codes, over time.

Take multicellular creatures like ourselves. We can be multicellular in large part because in the early stage of the development of creatures like us, replicating cells stuck together instead of just floating off into the ocean. Following their sticking together, the cells developed over time cell lineages which operated in parallel. This led to our complex selves. The sticking together arose from the way the cells related to each other – physical characteristics which had to do with the materials at hand. Now a significant amount of the coding of animal and plant genomes is connected to the way cells interrelate to each other within the organism.

The shapes and dynamics of the evolving organisms opened opportunities for further elaboration. Changes in the codes in the cells allowed those opportunities to be realized reliably, down over replication after replication of the cells and the organisms for which they coded.

Take the streamlined shape of large swimming creatures in the oceans, as compared with two and four legged animals on land. Streamlined shapes work better in a liquid, for a dynamic creature which moves itself in that liquid. Codes in the cells of those creatures which specified streamlined shapes of the body got the opportunity to reproduce more often than codes which did not, thus preserving the codes, as well as the type of creature coded for.

The dynamic processes of form generation and form, or relational, interactions, set the conditions for the coding, sustain the codes and give the codes operational effect. Codes and the organism carrying them are jointly on trial. Those codes – and those organisms -- which work with the shaping processes into which they fit replicate better than those which do not. Those which don't fit in, lose out in the game of life.

As Goodenough and Deacon point out, the life process differs from the simple massing of atoms into stars, planets, etc. Living organisms, using code strings to specify form and structure, build themselves into the surrounding relational settings. But the net effect is, when life is successful, form dynamically following -- and proliferating from -- form: form recreated anew iteration after iteration: correlation engendering correlation, through the life process.

8.5 Life Builds Groups

If we take the approach toward order building in the universe outlined here, we would predict that the life process would also form groups, and then groups of groups, in an hierarchy of life.

We might say that this hierarchy building would be the fundamental theme of life organization, as hierarchy building is a fundamental characteristic of the Universe.

And that is the case.

In life, we can see group building proceeding in two related ways.

One is the formation of groups which do not reproduce as groups. The group is formed from life elements, but the group of elements has a level and type of integration short of the group reproducing as a unit.

One example is the stromatolite cities, many of which were apparently made of single celled life units. Those stromatolites had and have huge numbers of denizens. This form of group living apparently got organized soon after life got started in the oceans.[38]

---

38 An entertaining, documented, well conceived, easily accessible and reasonably current discussion of stromatolites, and related structures, with proposed dates ranging back over three billion years ago, is available at the online FossilMuseum (n.d.)

Fig. 8.1: Stromatolite Cities.

Another example, separated by about three billion years of evolution, is human societies.

The other pathway is group integrations so tight that the entire group replicates as a unit. This includes eukaryotic cells, and the multicellular organisms which constitute almost all the life forms the unaided eye can see.

Two prominent investigators, Lynn Margules (1970), and, more extensively, Leo Buss (1987), developed a view of Earth's life structure as exhibiting tiered levels of (dynamic) aggregation of life units. In each level of aggregation, the component elements have mutually beneficial, or complementary, relationships.

In brief summary, the comprehensive Buss approach is cast in terms of replicating precursors which became inclusions in single celled organisms, thence the eukaryotic cell (which are, in Margulis' now widely adopted analysis, made up of single celled organisms), thence multicellular organisms, composed of eukaryotic cells. From there, we go to "social" organizations composed of multicellular organisms. Ants, termites and humans are prominent examples of social multicellers.

Commentators frequently display ambivalence about whether ants and termites have reached the next stage of integration, to have their hives recognized as "organisms" and "units of selection". I am inclined to favor the unit of selection point of view, taking into account the topological circularity of the hive generating the queen and consort, and the queen and consort generating the hive.

The great utility and appeal of organism form tracing and the gene-based tree of life depictions have tended to obscure this undeniable and fundamental group formation progression in the organization of life, and complexity in life. I have therefore assayed a complement to the "tree of life", which might be called the "layer cake of life". Please see the following.

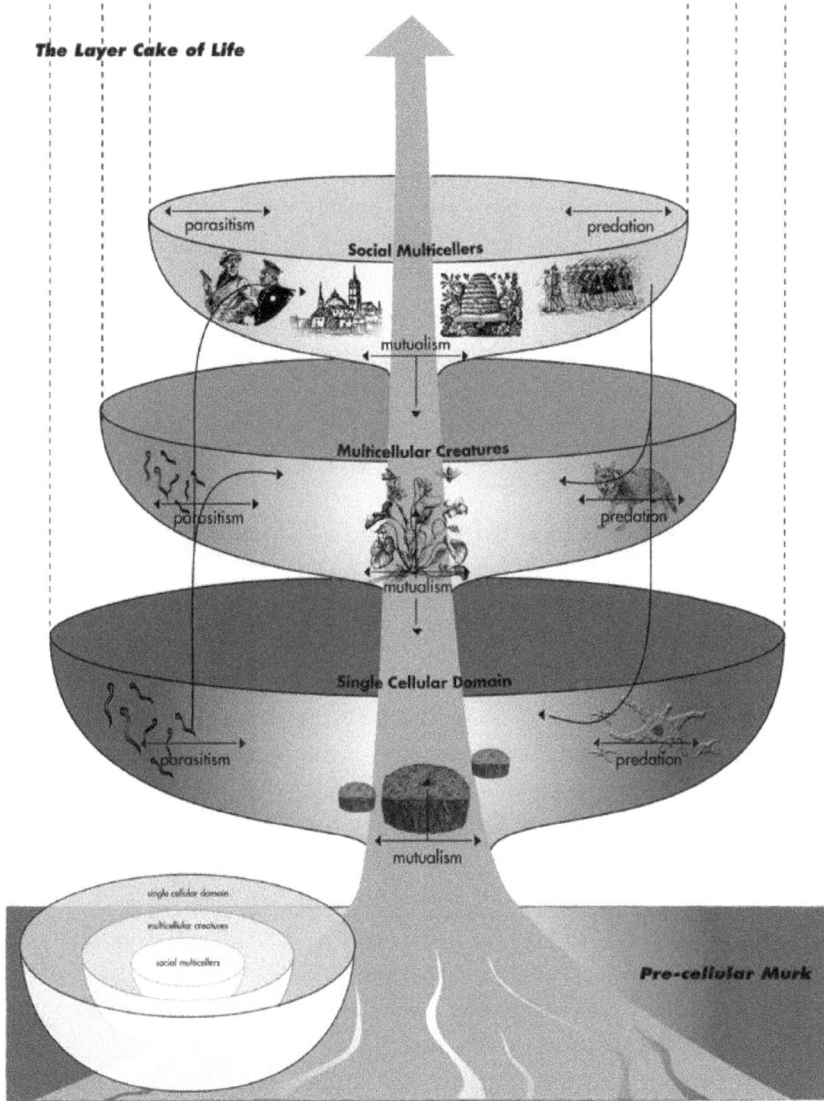

Figure 8.2: The Layer Cake of Life.

This chart has the virtue, among other things, of locating us humans in our proper place in life's organization chart.

We are a manifestation of group organization in large multicellular animals. Language is a necessary and prominent mechanism in our group organization. Biologists call animals fauna. So we might give ourselves the slightly ungainly acronym of LMMFG – language mediated mega faunal groupishness. (Well, it is better than having a big question mark on the T-shirt, and it is after all only five letters.)

We humans have created organizations at several scales. Our organizations take in energy and have what could be termed metabolism. However, though some of our corporate businesses demonstrate division and mergers, it is not clear that we have regularized replication of our social organizations. Perhaps if we establish self sustaining colonies on other planets we will have budded.

In the language humans use, group organization can lead to substantial energy efficiencies. These efficiencies provide ants, termites, bees and humans very good niches. Ants and termites have occupied their social-animal niche, at the smaller scale of insects, for over a hundred million years. They have come to fill up very substantial portions of biological space, so to speak. Their niche is large and apparently secure.

We humans are new in our niche, only a few hundred thousand years. We began to manage large scale exploitation of the niche only a few tens of thousands of years ago. However, we humans have no visible close competitors for our niche, and we can see the possibility of our niche also being large and secure. A portion of the remainder of this book will be about what we might do in and with this niche.

8.6 Evolution Creates "Mutual Information" Between Reproducing Units and Their Environment

"Mutual information" can to be taken to mean, in the life context here addressed, that each of two interacting elements, or systems, has a degree of order, or correlation, within it, and there is also some degree of correlation between the organization in one of the elements and the organization in the other element, or system.

Let us take the Earth when life started. It had organized and differentiated relational systems – solid rock with many distinguishable elements in it, liquid ocean with many distinguishable elements and compounds in it, a gaseous envelope, a regular rotation, a regular orbit around the sun, etc.

The first replicating biological entities had a significant degree of complexity, or organization in them also, though we do not know exactly how much. Those systems had to be compatible with, or to use familiar language, fit into, the conditions in the ocean on the Earth to maintain function and replicate.

But you might say the early organisms had it simple, if not easy, as compared with their descendants over time. Just lots of ocean and replicate away. But as the replicators began to populate the ocean, and later the land, they created conditions, threats, opportunities for any given replicator. There were things out there which could eat you, ally with you, be food for you, crowd you, jostle you, etc. Life got complicated. Probably quickly, in geological time. And continuously.

I have depicted a give and take sort of relationship between the replicators and their continually changing environment. The code strings in the replicators throws up a variety of combinations against the environment, sometimes winning, sometimes not. The winning combinations get replicated. In sexual replication, the winning combination then carries with it another round of variations.

As this process works out over time and replications, the code strings, and any given organism, come to have built into them a great deal of information about their environment. They are coded for day and night, for seasons, for camouflage to fool predators, for sociality, etc. There is a great deal of "if/then" coding. If the weather is cold, the codes for long hair are activated; or flight migration, etc.

In life systems with an observable high degree of internal information processing, and some range of action patterns, we can speak in terms of (somewhat) complex adaptive systems. This occurs within specific organisms, and in systems or groups of life units (like ants and humans).

We can observe adaptive computation in decentralized systems, like ants, within life units with complex centralized information processors, like humans, and in systems of "autonomous agents" like human societies.

Much of the information built into life units has accumulated in the slow feedback loops between environments and genome code strings. But with the evolution of animals with substantial (relative to other animals) processing power and complex language, we get "cultural evolution".

At one level, an individual life unit can "learn from experience", and formulate neural patterns which we can identify as "attention", "intention", "plan", etc. At another level, human language is a communal information construct, and has become a sort of communal cultural coding complex. This complex is now embodied in artifacts of many sorts, including but not limited to schools, universities, books, legal codes, organization manuals, data processing hardware and software, etc.

One might say that "science" seeks continually to amplify and refine the mutual information between our representation systems and the universe, using a set of inquiry, reportage, and verification protocols.

The human language system, combined with other coordinating elements in human society, and the spurt in energy use derived from tapping fossil fuels, has accelerated the development of "mutual information" between human organizations and their settings. That is, we have been rapidly multiplying our engagements with our surroundings.

Our pushy species has taken this to something of an extreme -- we are rapidly seeking to make over the Earth to make all of Earth our chosen niche.

However, we have begun to appreciate that this human-centered reorganization of the structures of the legacy ecosystem and humankind is producing effects within that legacy ecosystem which large portions of the legacy system are not geared to respond to, within their "natural", inherited, cycles of adjustment. Plants and other animals (but there is the possible exception of microbes) cannot always adjust to and respond to the changes brought by our expansion as rapidly as we have recently been able to cut them down, fence them out, and organize them to feed us.

As to the Earth situation, the only one we have access to close-up, one can think of the life process as a nested extension of order within in the regularities embodied in the sun/planet complex of the solar system. Here I have suggested that the state-space and state-potential search mechanism of the life process has produced a dense web of "mutual information" on the earth. This dense web of mutual information resides in and among the elements of the life system, as they go about their respective searches of the Earth system and each other.

The dense web of interconnections is self evident from our systematic investigation of earth life. The mapping of this back on the correlation mechanism may give us plausible schemes of "cause", in the sense of one thing following another, and, possibly, some additional help in figuring out how to operate in this framework. In a later section I will suggest some thoughts on how to orient ourselves to our situation, based on the prior-developed premises.

8.7 Evolution on Earth Builds a Set of Coexisting Relational Regimes Tending to Realize the Possibilities of the Life Process

Questions about where life is going in the Universe, what it is for, and what it tends toward have obviously been with us for a long time. As have been asserted answers, and frank speculations. I am going to put aside almost all this, confining myself to thoughts which have been closely tied to "science", and in recent decades.

One of the major questions is whether life -- as we humans experience it -- is "going" anywhere -- whether life on Earth has demonstrated any directionality. This debate was, in my opinion, definitively put to rest by Lenton and Watson's (2011) work "Revolutions That Made The Earth".

Lenton and Watson conclusively demonstrate several stages of life development leading to expansion of the scope, volume, variety and complexity of the current ecosystem. Since then, much other work has come to reflect the major expansion of the scope, energy embodiment, and complexity of life following the "great oxygenation" which was produced by photosynthetic organisms and life's expansion to land areas.

Eric Chaisson, Stephen Gould (1996) and numerous others have also addressed this topic.

Let us deal briefly and illustratively with some of the obvious facts "on the ground" -- and in the water and air. We can all see a number of major landmarks in life's evolution, which certainly suggest that over 3.5 billion years' life has spread, and developed complexity, in some recognizable ways. We see that life started out about 3.5 billion years ago pretty localized and simple, as compared to its manifestations today. With the development of photosynthesis, life gained a much broader energy base. With its extension to continental surface land it gained a much broader geographic base and greatly multiplied its mass. Insects, birds and bats took it into the atmosphere. It penetrates miles into the earth and around arctic ice masses. And it has developed complexity at least in adding levels of aggregation as noted in a preceding section.

In the late 1900s, Richard Dawkins (2016, memorial) was prominently identified with the view that the true utility function of life, that which is being maximized in the natural world, is DNA survival. Dawkins saw no guiding hand, so to speak, stewarding life toward a compositional maximization of a life "utility function". However, in my view that earlier framework – though highly interesting and explanatory in many ways -- has been shown to be far too limited in terms of the long term tendencies of the life process. It hardly begins to reflect the sense of the preceding paragraph, and Lenton and Watson's book.

I therefore have put Earth life in a broader context -- a thermodynamically propelled correlational process, which uses DNA and associated coding as a great convenience. The preceding section quoting from and referring to the work of Goodenough and Deacon spells out why I take this view, and how this view relates to Dawkins' theses.

Stuart Kauffman approached what life does from another set of perspectives. Most directly relevant here, he and associates worked out the concept of "fitness peaks" in the state spaces of life. A fitness peak, and valley, could be considered as the degree of fitness – survivability and replicability -- which might be reached by sets of code (read if you wish for convenience DNA codes). As the code elements changed, and as the codes in other life forms changed, the organism could get to higher peaks or slide into valleys.

An ecosystem might then be depicted as a fitness landscape with peaks and valleys. If an organism, or a species, was on a peak, it was fit and on top of the world – at least provisionally until the landscape changed around it. If it was in a valley, hard times or extinction. This is a conceptually elegant construct.

But Kauffman's construct potentially clashed with a problem posed by two of his associates, Bill Macready and David Wolpert. They devised a "no free lunch" mathematical construct, apparently quite sophisticated and sound, which rather unambiguously found that if one considered all the possibilities of fitness peaks and valleys, no search strategy – no mechanisms for seeking how one got to the peaks over the fitness landscape -- worked better than chance, or random peck and poke.

When I learned of this it caused no little pecking and poking in my own mind. What kind of strange non-random peak and set of peaks, more or less, had I, and all life, found, no better than randomly? Adrift in an ocean of randomness, what would become of us?

Kaufman found that his simulation of Earth life's mutation, recombination, and selection search system did well "on a special type of fitness landscape, where the high peaks tend to cluster and the sides of the peaks are reasonably smooth" (Investigations, p. 197). He went on to do some trial modeling and concluded that "coevolution by self-natural selection alone acting on individuals alone can tune landscape ruggedness so those landscapes are self-consistently well searched by the creatures searching them and their search mechanisms" (p. 205).

Here I offer a more generalized perspective, though it is not inconsistent with Kaufman's approaches. Life creatures do not face a "landscape" which contains all the possible combinations which mathematical notation can construct. Rather they face the power law patterns generated by the correlation mechanism operative in the universe. Each new trial combination of life, at each level of life organization, faces correlation-ordered relational structures which have a common statistical foundation. The system of recombinatorial trials, against a structured setting, followed by selection, tends to probe the set of "adjacent possibilities", in Kaufman's felicitous phrase.

Thus the thesis that this continual probing, and selection, have, over 3.5 billion years, built up a set of life state, realizations, containing much "mutual information", which we collectively recognize, as an assemblage, the ecosphere. If we consider life as a part of an order building process in the universe, we can say that it has worked off the order building mechanic in filling out some of the order building potentials of dynamic processes on and around the Earth's surface.

Chaisson's construct indicating increasing complexity, accompanied by increasing energy densities, is relevant here as well. However, as to the universe as a whole, I offer some qualifications.[39]

Relatedly, I, at least, have found no basis to say that earthlife has unlimited possibilities, or even a guaranteed future. Earthlife has taken some hard knocks – "iceball Earth" and several mass extinctions. The sun is scheduled to obliterate it a few billion years down the road, if Earthlife does not find some other venue.

Earthlife is dependent on the Earth's energy budget, and thus far on its own paltry harvest (less than one percent) of the radiation the sun showers upon Earth. As Smolin pointed out, thus far, unless humankind can substantially develop sustainable additional energy fixing technologies, carbon based life can manage to use only a very small sliver of local energy fluxes.

Also, I have not excluded the possibility that Earthlife may have long plateaus of stability in the over-all extent to which it realizes its scope and complexity; or even, on an aggregate scale, "punctuated equilibria".

---

39 Though Chaisson's illustrations are exceedingly instructive, some might, on first view, be taken to suggest the whole universe is getting more complex, throughout. If, however, we take into account the conditional probability structure of the universe and the power law distribution, we are, I suggest, forced into the conclusion that though over time more complexity seems to have been realized in the Universe, as Chaisson indicates, only small fractions of the universe have realized the high correlation, high energy density, high complexity possibilities which we humans are fortunate enough to inhabit. (I do not think Chaisson has explicitly indicated otherwise.)

Thus, this Section of this work asserts only that evolutionary life on Earth embodies a dynamic operational mechanic which tends to probe and to realize the possibilities of its local setting, and its decentralized, carbon based, bottom-up, hierarchical correlational program. One would expect much the same would be said of life anywhere else.

So I can offer the reader no projections about life taking over the universe, or ascending to some ethereal plane, or achieving whatever pinnacle humans might fancy. I will, however, in a later section, attempt to address what we may need to do to have a more ample ecosystem making more room for us, here on Earth and nearby.

Let us now approach the particular question of whether Earthlife is becoming more complex over evolutionary time, or evolutionary process, rather than just "bulking up", so to speak.

People have addressed this in a variety of ways, of course. Among the approaches have been to try to see if fossils show visible elaboration over time, and to count the number of species of plants or animals in a defined ecosystem setting at different eras.

Bob Ulanowicz's detailed canvases of ecosystems provide a way of mapping ecosystem energy flow and differentiation, among other things, using information theoretic tools. Such canvases could be extended to economic systems, and of course to other systems, with, in my view, probable utility. [40]

In his book "Full House" Stephen Gould(1996) created an interesting way of arguing that the life process has not necessarily embodied an increase in average complexity in the biosphere, though an over-all increase in the biosphere over time could lead to a corresponding increase in the total amount of complexity in the ecosystem. In the course of his argument he suggested that if the complexity of organisms were arrayed on a right angle diagram, the far-right statistical tail of complex organisms would be increased in over all size over evolutionary time if the total mass and volume of the ecosystem increased over time, without reference to whether average organism complexity increased.

In the last chapter I will adapt Gould's graphical construct and combine it with other concepts introduced thus far -- largely, the ubiquity of power laws which set the slope of the complexity curve -- to suggest a definition of an approach we would be required to follow if we want sustainably to increase human welfare over time in our ecosystem.[41]

---

40 Ulanowicz's analyses would be a considerable upgrade over the older input/output tables of V. Leontief, though these were an analytic advance in their day.

41 Using the power law curve on logarithmic axes, one of which would use mass measures, might also be a path to quantifying the total level of complexity at a given degree, in the Universe, were we able to quantify, reliably, complexity measures, and relate them to mass measures.

At this time, I would suggest that Chaisson's approach to complexity would often provide an highly convenient way to measure the extent to which life on Earth has realized greater areas, or situations, of complexity, within its over-all scope, over time, on this Earth. Also, his approach helps relate that question to determinates of complexity, so as to help guide us in attempting to intervene in our immediate and larger circumstances to make more room for maintaining and expanding the complex things we like to be and to do.

For example, the complexity of an individual creature, or complex of creatures, can be related to the free energy density in the creature or complex of creatures. A similar approach can be taken to human artifacts, like cities, computers, etc.

As to the biosphere, well informed specialists might assay the masses, energy budgets, and energy densities of ecosystems in various life eras, and in various subsets of the current biosphere.

So one can not assert that each element of Earthlife has generated within itself more and more complexity, steadily (except for mass extinctions) and inexorably, over time -- or at least I would not.

But we ourselves are evidence that after a long string of life innovations (photosynthesis, multicellular life, land based life, flowering plants, etc.) life has engendered a significant innovation in large scale (billions of units) socialization of large animals, and this is another step in life's hierarchy building. This socialization, or group-making, process has generated an increase in the complexity of life realizations.

Chaisson calculates that human societies are more complex – have higher energy densities – than the megafauna (us) who make them up, and our computer chip artifacts have higher energy density rates still. (Chaisson, pp.158, 198, 200-207). In Chaisson's calculations, humans have achieved, or one might say embody, an energy density and complexity peak. This will be graphically illustrated in the last Chapter of this monograph.

One can project that this life innovation, while threatening to do considerable damage to the ecosystem which created it, may, if we are fortunate and use well the tools given us, have the potential to increase the scope and complexity of Earthlife -- to add to rather than subtract from the Universe's order construction regime.

This product of evolution -- humans -- has created a new source of energy flow through the ecosphere which may substantially alter the over all landscape of evolutionary products -- i.e. substantial energy flows other than those arising from current biological photosynthesis. In the world of the 21st century CE, or about 10 millennia after the onset of systematic human harvesting of current photosynthesis products, the products of photosynthesis in prior geologic eras have been tapped systematically and at large scale (about 10% of the current photosynthetic yield). Humans have used this resource to create an "industrial" set of systems which service human desires apart from, though complementary to, the harvested biological energy flow. We now have an industrial metabolism complementing the metabolisms of individual life forms, and the collective photosynthesis-derived metabolism of life to date.

This has resulted in a great proliferation of humans, crowding down or out of a number of large animal species, alteration of much of the vegetative portion of the ecosystem, and possible significant heating of the Earth atmosphere, with prospective resultant additional significant alteration of the shape and content the multicellular ecosphere.

If humans cannot replace this "fossil fuel" energy flow with sustainable "renewable", or current, or daily, energy flows at current scale, the human alteration of the ecosphere may eventuate in approximately the types and levels of the "agricultural" era just prior to the "industrial" era.

If humans can replace fossil fuels at scale, sustainably (probably largely by wind energy harvesting and artificial photosynthesis), then life evolution on Earth will come to work off two major energy flows, rather than one.

This might have implications for life structure on earth of a magnitude similar to the introduction of photosynthesis itself. (Though I would not be so hopeful, or perhaps rash, as to predict that we will ever reach the total level of biological photosynthesis.)

I do not think this potential evolution will alter the basics of the evolutionary process outlined previously in this Chapter. However, it would seem likely, if it occurs, to alter the patterns of species, plant and animal, in the ecosphere, and possibly substantially to alter the visible landscape of the Earth on which it occurs.

This is now, and would be in the future, a major organizational challenge to humankind. I will attempt to address some of the implications of this challenge in later Chapters,

This has been a rather fast gallop through the life process. We have gone through a basic schematic definition of life to means of assessing the total complexity of life on Earth, and prefaced a potential alteration in the scope of the life process on Earth, in less than 30 pages. Let us see if this compressed characterization of some of the salient features of the life process helps us address pressing issues of the day, from the perspective of humans like us. Let us therefore now proceed to a Chapter on energy, or energy flows, which of course have shaped and will shape evolution to date and in the future.

9 Energy

We all know that energy flows are prerequisite to life. Preceding chapters have shown that the amount of energy which living things can access and incorporate (presuming for the moment that the energy level is sustained and consistent over time), determine how much life can exist and how complex it can become.

Let us set a frame for considering the energy flow aspects of life on Earth.

First, there is a great deal more energy in our neighborhood than Earthlife now accesses.

In the 1970's Likens and Whittaker (1975) estimated life's gross energy capture by photosynthesis at @5.8 $x10^{21}$ joules/year. The total solar energy flux on the Earth's surface is currently estimated at about $5.7x10^{24}$ joules/year, according to the United Nations Food and Agriculture Organization.[42] Thus life manages to capture less than one percent of the total radiant energy flux on earth, in visible light wavelengths. On this sip from the Sun's bounty rests the vast bulk of surface Earthlife, the more complex part of life here.

Roughly half of the gross energy capture is used in the respiration of the capturing organisms -- just moving the machinery. An estimated half goes to "net primary production" [43] -- the production of biomass.

If we put this in earth-size perspective, the annual primary production of earth surface biomass is about $1.6x10^{11}$ metric tons, leading to a total standing biomass of about $1.8x10^{12}$ metric tons.[44] That is about $1/3,000,000,000$ of Earth's $5.972x10^{21}$ metric tons of mass.

At first look, humans appear to account for a pretty small proportion of Earthlife's energy budget as invested in biomass. If we estimate human total biomass at $5.4x10^8$ metric tons (about 60 kilograms (10^3 grams) per person times 9 billion ($9x10^9$) people), then we are less than one half of one percent of total earth surface biomass.

If we think in terms of the energy budget we feed off, or entrain, the numbers look a little more imposing. Let's think, for the moment, of energy budgets in terms of the proportion we consume of the biomass which is created by life's current energy cycling.

---

42 United Nations Food And Agriculture Organization, retrieved at http://www.fao.org/docrep/w7241e/w7241e06.htm#2.1%20photosynthetic%20capture%20of%20solar%20energy

43 Likens and Whittaker, Supra, at p. 310.

44 An organization in the UK called Globalnet has designed a very convenient Web presentation, retrieved at http://www.users.globalnet.co.uk/~mfogg/icons/calc3.html , which is based on data such as that compiled by the National Aeronautics and Space Agency, retrieved at https://daac.ornl.gov/NPP/npp_home.shtml. One can also see Wikipedia, which does have extensive annotation, at http://en.wikipedia.org/wiki/Solar_energy#cite_note-9.

Humans directly consume as food less than one half of one percent of the net plant biomass per year, according to Peter Vitousek et al (1986).

When one counts other activities, including the consumption of our domesticated animals, we cut a wider swath. Vitousek has calculated that when one counts in wood harvests, firewood use, what our domesticated animals eat, etc., we consume about three percent of the biomass created each year.

But now we come to the shocker which Vitousek and associates calculated. They conceived of seeing our place in the biosphere in terms of what we have "co-opted" in it. In precise language, he initially defined this as

"material that humans use directly or that is used in human-dominated ecosystems by communities of organisms different from those in corresponding natural ecosystems". (p. 370).

Following the 1986 publication of his first set of calculations, a continuing set of reviews has been mounted, under the general title of human appropriation of net primary productivity on earths land surfaces ("HANPP"). Vitousek and followers are talking in terms of cropland, pastureland, or grazing land, and altered or harvested forest land -- places where we have harvested, killed trees not taken, imposed shifting cultivation, converted to tree plantations, and done land clearing -- all plus areas of human habitation and some aquatic harvesting.

As a metaphor, we're talking about the portion of the ecosystem we have converted into a funnel for our appetites and activities.

Vitousek et al argue that we have "co-opted", or made into a human-serving funnel, @ 30-35% of the net primary productivity of the biosphere on land.[45] He got higher figures when he counted the net primary productivity foregone by reason of human activities – that is, counting potential productivity precluded, we affected about 40% of net primary productivity of photosynthesis on Earth/

Vitousek and friends appear to be shocked, concerned, appalled.

I am impressed. That's a pretty ingenious feat we humans have pulled off. However, the amount of net productivity which we may reduce, in this process, is also a major consideration.

And you can understand why some would get nervous about whether Mother Earth is going to keep on letting us get away with it.

However, in tapping fossil fuels, the energy trove of life past, we've gone beyond shaping life into a funnel for our own energy flows. We've been adding to life's current energy budget.

---

45 Vitousek's estimate has some heroism in it. We are talking of large quantities, and rough, or approximate, means of measuring. But I have not found a serious challenge made to it since it was published. The estimate is not inconsistent with the common perception that we have cut down and use a lot of forest, created a lot of crop and pastureland, and fish the seas pretty actively.

In the second decade of the 21st century the human species generates in commercial and industrial uses upward of 5 x 10 ^{20} joules/year.[46] We get most of this from fossil fuels, nuclear energy and waterfall management. The largest sources are fossil fuels and nuclear energy.

Thus human-created energy flows now approach @10% of total energy flows which the photo synthesizers of Earth manage.

This may not sound like a lot. But if you bring into focus that this 8-10% is concentrated on less than 1% of the biomass, you see the privileged position we have carved out for ourselves, and the source of our impact on the life community on Earth.

We have bigger plans, of course. Some projections of future human global energy use in the latter part of the 21st century have tended to fall in the range of three to four times current amounts -- 1.1 to 1.5 x10 ^{21} joules/ year, or 1100 to 1500 exajoules per year.[47] (Though at the time of this writing, most entities engaged in energy use projections tend to focus on the 2030-2050 time range, as this is perceived to be a critical time period for the substitution of renewable energy sources for fossil fuel sources, at scale.)

If we were able to get to doubling or tripling current artifactual energy use levels -- which would tend to be desired by the very large populations of India, China, and Africa -- humans would be adding @20-25% to the energy production of life on earth, or displacing that amount of "natural" energy flows, or some combination of additions and subtractions.

Humans are, at least momentarily, enormously advantaged among other life forms in energy consumption. But we are also precariously positioned.

We all know intuitively that our present high standards of living come from those fossil energy flows – working off the stored energy in the remains of long dead organisms. But some time ago a group of investigators working with techniques developed by H.T. Odum (1983) quantified this, in an interesting, thorough, and, if you are paying attention, very sobering way.

Odum and his collaborators have, among a number of other very interesting things, compiled analyses of the energy budgets of many of the major nations. In these analyses Odum et al attempted to calculate the contribution of energy from all significant sources to the energy flow through the human system – like sunlight, wind, tides, waves, topsoil formation, oil, nuclear generation, etc.

---

46 One can get data series on this, on a continuing basis, to allow updating, from the United States Energy Information Agency, the International Energy Administration, and the World Resources Institute, among a wide variety of other sources.

47 One can get a number of energy estimates from a number of sources. These estimates differ widely over time. The range noted is not uncommon, but is also greeted with some skepticism. See e.g. Smil, (2005), Chapter 3.

With exhaustive analyses, the scope of which I have barely hinted, Odum's group was able to calculate the ratio of the then-existing populations which could be sustained at then current living standards on renewable energy sources alone, if we did not use fossil fuels. (Odum et al did not count nuclear energy as renewable. And he did not estimate that solar energy or wind energy technologies would yield large returns. I will come back to that.)

On the basis of the technologies available at the time of his analyses, Odum et al estimated that in the advanced United States economy, renewable energy sources could provide only 12% of the energy flow we then obtained largely from fossil fuels. That is, we could sustain only 12% of our then-current population at current energy consumption levels on the biological and traditional – pre-industrial—energy flows. Other highly industrialized, wealthy nations were not much different.

To drive home the point, if we had only the energy sources which Odum counted as renewable, or long run sustainable, in the United States, we could, he calculated, sustain only about 1/8 of the people we had at then-current levels of energy usage, or the same number of people at 1/8 the energy usage.[48]

Economic conditions have changed somewhat since Odum's work was done. Our per capita energy use is substantially greater – which makes the energy use reduction ratio greater. However, we now can see our way clear, we think and hope, to tapping solar energy and wind energy at large scales.

Fossil fuel supplies are finite, so sooner or later the humans on Earth must either cut back on energy use, or create new energy supplies to substitute for the fossil fuel bonanza of recent centuries, or some combination of the two. It seems unlikely at this time that we could squeeze another 7-10% of the energy flow of the legacy system ecosystem out of it without damaging it.

At this time, the combination of proliferating development of human societies and the side effect of accumulating "greenhouse gases" (largely carbon dioxide) in the Earth's atmosphere has presented human society with a difficult issue.

It appears that if we were able to continue to mine "fossil fuels" -- hydrocarbons remaining from earlier eons of life development -- at the rates we now do, and with recent rates of increase, we would raise the temperature of the atmosphere to such an extent as to melt some or all of the Earth's icecaps. Organizations such as the Climate Emergency Institute project the possibility of 'runaway' scenarios which would eliminate human civilization, and drastically alter the ecosphere as a whole.

The geographic areas possibly inundated in some of the more moderate projections now support over a billion people and represent a very large investment of human capital in cities and agricultural areas (for example, the largest cities on the

---

48 Obviously if we did not have the fossil fuel flows or an approximately equally effective set of successor technologies, we might or might not end up with either 12% of the people surviving or all surviving at 12% of the energy throughput each. This is just a shorthand way of comparing quantities of energy and people in a current setting.

East Coast of the United States, and prime agricultural land there, as well as Northern Europe and much of Russia).

In addition, scientific modeling -- not perfect, of course, but informative -- projects alteration of weather patterns, vegetation patterns, and habitation zones to such an extent as significantly to disrupt existing growing seasons, agricultural productivity, and global ecological carrying capacity.

As of this writing, the nations of the Earth have begun an effort to limit the emissions of greenhouse gases in this century to prevent major sea level rises, and resulting prejudice to land and ocean biological carrying capacity. This effort has brought some participants to the clear and stark realization that the rapid rise of human populations and humans' industrial metabolism, so to speak, over the last two centuries, has been built on the use of the life-residue hydrocarbons, AKA fossil fuels.

That has brought more clearly into focus the realization that curtailing such fossil fuel use, without replacement energy flows, would limit future growth of the human enterprise, or, in the extreme, greatly shrink human civilization and the quality, or enjoyability, of human life.

It appears to me that much of the "climate change denier" reaction to the scientists' work arises directly from this "gut" realization, or feeling. This is bitter medicine. Many would prefer not to recognize any need to cut back on fossil fuel use.

However, many are willing to credit the work of dedicated and competent professionals in climate modelling, and consider what needs to be done to prevent major alterations in our living conditions on this globe. This has led to efforts, around the globe, to accelerate the development of "sustainable" energy production systems. Such systems include solar energy technologies, wind, geothermal, tidal, and other sources. Also, nuclear technologies could play a significant role, if they can be further developed without major fossil fuel greenhouse gas emissions.

This focus on developing non-fossil energy supply systems has, in only a few years, led to the realization that rebuilding the basic energy support system for "industrial" civilization will take decades, if not more than a century._ Vaclav Smil (2005) has been constructively cautious in this regard. In the process, we will need new technologies (such as new or newly expanded energy storage systems, and conversion of solar energy into "chemical" fuels).

In plain language, we now can see more clearly that we face a make-or-break transition in our human civilization. We will have to do a ground-up renovation of our civilizational house of cards, while inhabiting it.

In the climate control issue we have the opening stage of a test of how we will cope with this transition. One could take the perspective that the atmospheric temperature control problem has forced humans to focus sooner than they otherwise might have on the larger transition from fossil fuels to long term sustainable fuel. This is the larger and fundamentally defining issue for human civilization.

That longer term issue -- whether forced by "climate change" or not -- is whether we can continue to grow our now-massive global population, increase the living standards of humans globally, and further our many technological and scientific arts, or must face a cap on our capacities, or even a massive world wide reduction in all these things, including a reduction in the integrity of governance systems which often may not able to cope with the stresses involved.

In a way, however, the cautions suggested by "climate change" concerns may be a boon, or blessing (even if some consider it a well disguised blessing). If, impelled by the "climate change" problem, we can use fossil fuels to lay the foundations and for and erect the frame of "sustainable" mass energy supplies globally before we face the low energy rates of return on energy invested in fossil fuels in the tailing-out process -- at the near-exhaustion stage of these supplies -- we may be spared an even more difficult situation.

That more difficult situation would be a quandary, or dead end, in which investment in fossil fuels yields low energy returns forcing reduction in our civilization before sustainable energy technologies, effectively integrated into civilization support systems, have reached levels of maturity providing a sufficiently high volume and high rate of return of energy on energy invested in them to support our industrial civilization at or near levels we would find accustomed and comfortable.

In other words, more briefly, we may have time to take care of the energy transition problem before it becomes unmanageable, and that unmanageability disposes of most of us humans. That is, if we look ahead and act with some foresight and discipline.[49]

Many people fervently espouse cutting back on energy use. Underlying this may be a deeply held assumption that we will not in the long run be able to improve on the energy yield of the legacy ecosystem, coupled with an aversion to damaging the system which created us and has thus far sustained us.

I am strongly inclined to favor increasing the long term energy supply for Earthlife. In my view it would be a great triumph of evolution if its large-animal socialization innovation -- us --were to open up new energy windows, and thus to increase the scope and complexity of Earthlife.

Beyond my own preference, there are now several billions of humans who prefer that they and their descendants live more ample lives, rather than diminished lives, or no lives at all. And underlying all this is the basic dynamic of life set out in the prior chapter. Life is built to capture and incorporate energy. That is "what life does", as we humans would say.

---

49 You, and Hollywood, can imagine what life for our progeny could be like if only, say, one billion of our current @8-9 billion people could survive - that is, each of your progeny faced only a 1/9 chance of survival. The Earth would survive. We are only one species in it. But humans would be unpleasantly humbled.

Let us briefly recapitulate why enlarging the human, and earthlife, energy budget would appear to be technically possible. As a thorough study on energy technologies by an International Panel on Climate Change (2011-2012), a seminal report by Tsao, Lewis and Crabtree (2006) at Sandia Laboratories, and other materials have documented, solar energy impinging on the surface of the earth per year (@5.5×10^{24} Joules) dwarf the biological photosynthetic energy intake. While solar energy is diffuse, artifactual solar systems currently extract @10 times more of the light energy striking them than do biological systems. The costs of artifactual solar energy, dropping for decades, have begun to approach fossil fuel electricity generation costs, and currently seem to be decreasing about 20% with each doubling of installed capacity.

This is an evolving picture, with numerous initiatives, widely discussed on a continuing basis. Energy yield ratios at this level are not as good as were early fossil fuel ratios, but with wise management, efficient energy transport and storage systems, and slow growth in populations may be sufficient for high energy, high complexity civilization.

As the Intergovernmental Panel on Climate Change comprehensive study of energy sources and the Sandia summary establish, wind and solar energy technologies are technically capable of generating more electricity than the world consumes, by several multiples. But complementary technologies are needed. Research on energy storage systems of various sorts proceeds apace. As the Sandia study suggests, if and as solar electricity and wind generation costs approach 1-2 cents per kilowatt hour, chemical fuels may become producible in large quantities, at costs which would compete with current fuel technologies.

Now underway is a worldwide set of searches for options to complement solar and wind technologies. Candidates include a variety of large scale electricity storage systems, household sized storage systems, use of electric vehicle batteries as electricity grid complements, nuclear energy installations of various types, the use of long distance transmission facilities to mix and match geographically and technically diverse energy generation and storage systems, extensive use of nuclear technologies (though perhaps along the lines of small nuclear reactors and thorium (rather than uranium) based processes), and much more.

Rather than attempt to summarize and project a complex, shifting, and exploratory set of research and development activities in this overview of energy driven ordering possibilities, I choose to alert readers to the many sources of information constantly available to interested and literate persons, and focus on two major sets of issues presented by the energy transition now before us.

One set of issues relates to making a successful transition to large scale energy supply for human civilization, and the other to how to integrate the additional artifactual energy flow and the legacy photosynthetic energy system which sustains most of Earthlife.

As to the second set of issues, humankind would be altering Earthlife from a single major energy supply system to, or toward, a permanent (as long as we are around and well organized) dual energy supply system. This raises the prospect of "crowding out" some of the legacy system, and, more generally, the challenge of managing additive integration of the two energy streams, as distinguished from destructive interference between them.

This problem, or issue, is made even more challenging because both energy supply systems are and will be complex, we have only begun to understand the interdependence dynamics of the legacy system, we do not completely understand the energy system we have been building, we find it difficult to predict the results of our actions, and we have major problems and challenges in composing and coordinating the organizational systems and ambitions of our recently-globalized human social complex.

We have to be able to coordinate human activities to be able to get more action out of the universe than we spend putting into it. Earthlife does this, but has had about 3-4 billion years to work on the trick. Earthlife worked at microscopic scale and masses. We are moving around macroscopic factors of production. We envisage a few hundred critical years.

Humankind has thus far done well in tapping the Earthlife energy stockpile on a global basis. That was in a sense the "low hanging fruit." Making up new energy supply arrangements seems likely to be more difficult.

In approaching this area, I suggest consideration of two very different perspectives. One involves very considerable computations, having to do with how to discern energy supply options, and choose among them. The second assumes all this work is done, and fits the question of energy and civilization growth potentials into a physical model of civilizational access to energy supplies. This latter construct can be used to address potential effects on the environment which might compromise the entire process of amplifying energy flows on earth by artifactual means.

Let us proceed to the first approach. As a considerable literature has put the point, a critical question is how much energy return on energy investment – EROEI -- we can manage. Several analyses put current estimates of EROEIs for major "alternative energy" technologies ranging from 5/1 to 20/1. Energy returns for oil and coal are put, generally, between 15 and 30 to one. The returns were often higher at earlier stages of fossil fuel recovery. See, e.g. Hall, et al (2014)

These EROEI ratios for renewables have a direct relationship to the energy surplus, or free(d) energy, available to humans. If the EROEI of renewables are low relative to historical fossil fuel returns, we would see smaller energy surpluses, for any given quantity of capital and energy input dedicated to energy production.[50]

---

50 One might project higher ratios in the future, as a result of "the learning curve" – the frequently observed increases in productivity of new production technologies over time. In the relational systems language I have used, one might depict such developments as the result of a combination of "mutual information" between the production process and its supply, distribution and use arrangements, and paring of the process to the minimum (lowest energy cost) embodiments needed to perform the specialized function involved.

All other things equal, this would result in lower standards of living, or slower gains in standards of living, for humanity. On the other hand, if future renewable EROEI were better than future fossil EROEI, all factors considered, this would be the path we would be compelled to take.

One can express this situation and its relationship to what most concerns us -- human welfare -- in a direct, quantitative and sweeping way. Assuming the EROEI is positive, a crude approximation of per capita welfare[51] for a citizen of an artifactual energy civilization would result from iterative operation of a formula such as

$$EPC = \frac{(EK * EROEI)\text{-}Er\text{-}Ed\text{-}Eenv}{P} \qquad (9.1)$$

Where EPC is per capita energy availability, EK is a scaling factor composed of energy going through a "capital" structure specialized for production of energy, Er is the energy funneled back into the production mechanism (for maintenance and investment), Ed is the energy dissipated from the civilizational structure, Eenv would represent any energy load on the civilizational structure by harm to the environment in which it functions, and P is population.

The linear presentation of this formula suggests a snapshot. However, the system would be cyclic -- have an energy producing apparatus, vis a vis human, input some energy into it, reap and distribute the energy obtained, reinvest some energy, employ the remainder to other human endeavors, and keep cranking.

I have installed the EK factor as an identified component for two reasons. One is to envisage the process as starting with a given capital stock at a given point in time. This allows modelling to make explicit the relationship of the EK scaling factor to the population.

Using a model such as this, one can imagine differing endowments in different polities -- e.g. higher or lower current capital endowments, higher or lower population

---

We have been seeing "learning curves" for major new energy supply technologies. We can observe that the EROEIs for fossil fuels started off high, and "renewable" technologies have been narrowing the gap, as of this writing.

51 This approach assumes that human welfare correlates with energy availability to the population. A number of surveys indicate that the correlation, in today's world, is not linear – not one to one. See e.g. Bent, Orr and Baker, Ed. (2001) The surveys seem to suggest, taken together, that at low energy production levels, the correlation is very high, but at energy production levels prevalent in the most wealthy countries today several measures of human welfare and happiness have not increased proportionately with higher energy use. I am assuming energy use in the range up to per capita levels which produce good measures of health, longevity, education, productivity, and psychological equanimity. Beyond this the correlation with "welfare" as thus defined may break down, but the human recipients of the energy flow will presumably still consider themselves advantaged. One might also assume a "sustainability" constraint.

counts, higher or lower EROEI results. One can also investigate the effects of postulated increase or decrease as to each of the three factors -- population, propensity to reinvest in capital, EROEI, etc. Some interesting possible relationships are noted in the footnote.[52]

When considering the iterative operation of the investment cycle, one could at any point calculate the expected EROEI over time of each of an array of technologies.

If, for example, fossil fuel real energy costs were to increase over time, while solar and/or wind costs, relative to energy production, were reasonably and accurately expected to decrease over time, this could be factored into the investment allocation at any given point in the cycle.

Something like that seems to be happening at present. The general public has been providing "subsidies" for renewable energy technologies. In effect these devices are ways of steering investments into technologies thought likely to produce better energy returns on energy invested in the future (considering the energy costs of all identifiable factors, including "externalities" such as air quality decreases and/or global warming).

This sort of calculation could be devised for various polities, local and regional groupings, and the global civilization. The question is common across a broad range of settings and technologies.

---

52 One could do combinations of variables of varying magnitudes -- e.g. large population, low capital, high or low EROEI; small population, a standard or average per capita capital factor, and an high EROEI; populations with differing allocations of energy flow back to capital (in some current parlance, "savings"), and so forth.

One can elaborate, or disaggregate, the reinvestment process along these lines. Imagine N capital sectors with varying EROEI at a given time period --- $k1, k2, ...kn$. Allocate reinvestment according to its EREOI rank (this could be done by market or central allocation mechanism (within a firm or society).) Total reinvestment =SUM Reinv $k1, k2, ...kn$. (The notation used in this para is altered from standard mathematical notation, slightly, to ease publication typography and facilitate reading by non-technical persons.)

Now suppose that one can project gains (or losses) in EROEI at differing rates among the energy generation sectors over consecutive cycles. One could, additionally, alter the order of reinvestments by a factor, as to each, scaled to the differences in anticipated EROEI gains (or losses) as to each sector over time (cycles).

In effect, we now do this, roughly, by giving tax credits for investments in selected sectors, and letting the investment markets then respond to the resulting weighted returns. This device is designed to accelerate the path to economies, or EROEI gains. It assumes past rates of gains, as in photovoltaics and wind energy, can be expected to continue and produce large efficiency gains over future time periods. The Bloomberg report cited herein makes this assumption as to photovoltaics and wind energy, over coming decades, and there is a good deal of experience and current activity to support this assumption.

(Note that the bias, or acceleration, given the cost-decreasing sector withdraws energy from other sectors otherwise allocated to them strictly on the current cycle EROEI rankings. Acceleration has its costs, as always, and they are often complained about in discussions of energy public policy.)

If our social systems do not break down from the strains of needing and creating new energy supplies, humans will be addressing questions of this sort for at least a couple of centuries or so. (If our social systems do break down, such humans as may survive will be addressing the consequences of that for at least two, and certainly more, centuries.)

The following graphic depicts a situation we seem to be encountering, with likely attendant social stresses. The graphic relates to the energy transition issue decreasing EROEI as to fossil fuels from levels unlikely to be obtained from "renewable" energy technologies, and increasing EROEI for solar and wind energy sources (as well as ancillary technologies such as energy storage). This has been the case in recent decades.[53]

The schematically presented intersection point between fossil and renewable EROEI is juxtaposed with posited global (or local) EROEI levels needed to sustain an industrial civilization of roughly the current sort.

I have projected a range of EROEI levels because of the uncertainty evident in current calculations of globally required returns on energy generating investments, and some uncertainty about what global polities are willing and able to 'settle for', and sustain, as to standards of living.

I depicted an intersection of fossil and renewable EROEI which falls within the postulated feasibility boundaries. Current technologies, and their development paths, suggest that this may be feasible. Just about all humans aware of the issues of 'development' either assume or hope that things will work out this way. But in my opinion the universe does not guarantee us this result.

The left to right downward sloping line depicts decreasing EROEI rates from fossil fuel sources. The left to right upward sloping line depicts increasing EROEI for renewable energy facilities. The two dashed lines depict two alternative EROEI levels required to support civilization at something like its current scale and energy throughput.

---

53 A paper by a group headed by A.S. Hall (2014) has developed documentation for the assumptions that fossil fuel EROEI are falling and those of renewables have recently reached levels higher than in previous decades. A recent paper surveying a host of researches on solar energy EROEI reports levels roughly estimated at between 8/1 and 30/1 for different technologies. Bhandari, et al (2015). A paper by an independent commentator uses these data and others to compare the EROEI of various means of generating electricity and puts various solar technologies in the 10-25/1 range, and rising. See Naam, R. (2015)

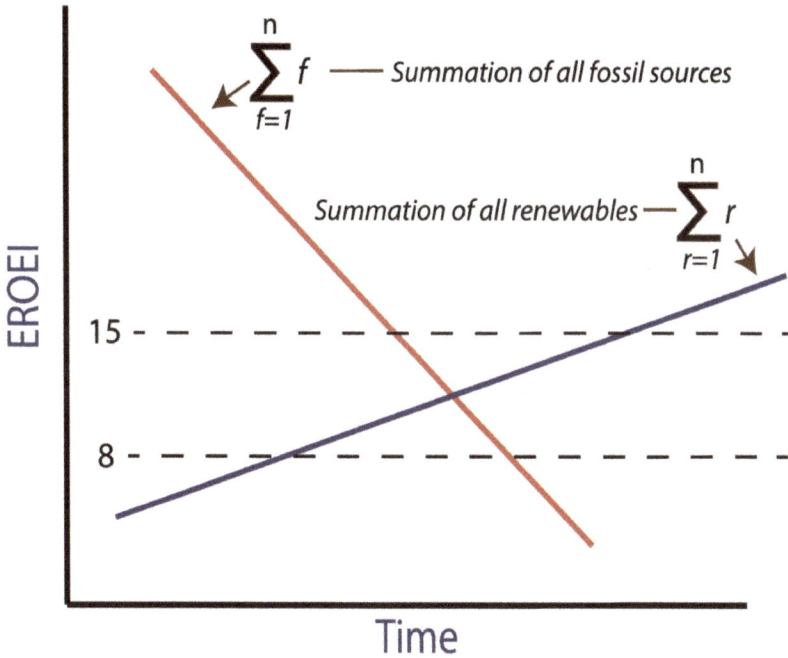

Fig. 9.1: Energy Source Transitions and Civilizational Support Boundaries.

As one can see, the net aggregate energy surplus available to human society decreases until the renewable technologies (hopefully) become available at higher levels of EROEI. The slope of the curves depicted presumes that "renewables" EROEI are unlikely to reach the rates seen in the earlier stages of fossil fuel exploitation.

As noted above, this hypothesized placement of the intersection of the EROEI curves and the placement of hypothesized levels needed to sustain a typically modern industrial society are designed to point out that the energy transition ahead could lead to either a diminution of growth levels for a period (and thereafter, very probably, in growth levels moderate compared to early "industrial age" levels), or even a decrease in the civilization level as a whole.

This latter outcome would mean decreased gross domestic products, smaller houses, poorer diets, lower education levels, shorter lives, and so forth, in the currently more industrialized parts of the world.[54]

---

54 Implicit in this form of analysis is the possibility that a civilization might hit a stasis point -- that is, the EROEI at the point of energy supply facilities would just suffice to sustain a civilization at its then current form of development, given dissipative energy losses and need to maintain a supporting biological ecosystem, and return enough energy reinvestment to keep the system cycling. This would imply a static civilization, with little innovation and little or no perceptible "progress". Some agricultural societies at some points in history may have presented this appearance.

And this sort of situation would, of course, lead to enormous stresses on current national and international governance systems, particularly if the citizens in the affected civilizations did not understand the processes they were going through. Pearce (2016) has elaborated on this aspect of the situation in an International Affairs Forum publication.

In my view, data currently on hand as to energy production technologies and the energy demands of the current world population warrants a placement of the curves at the levels and intersection points generally suggested by this diagram, justifying the implications suggested in this text. Work in progress by an host of researchers on a variety of issues subsumed by this scheme will, hopefully, better inform our progress and our prospects.

I have thus far characterized this form of EROEI analysis in global terms. However, it could be used as to individual nations, or regions, or other frameworks. This might be useful if differences in endowments and opportunities were to vary substantially among polities and data could be marshalled with sufficient discrimination.

The analysis could also be used with variations and additions. For example, one could calculate the scope-wide EROEI at any given point as the weighted average of the sectoral EROEIs, and calculate, or project, whether the differential between sectoral EROEIs over time were constant or might vary.

Though I have suggested various uses of EROEI calculations, this approach is attended by significant issues.

In the formula set out above, I have suggested that one explicitly recognize the energy going back into the energy production facilities, in maintaining them, and 'turning the crank', so to speak. But this leads one into labor costs, which entails some support for individuals in the society. That could be seen as energy available to those persons.

Trying to account for, and to allocate, the general maintenance costs for civilization as a whole or in a sector could be challenging. Accounting for environmental degradation might be handled in more than one way, and is often contentious.

Other significant issues have arisen. Practitioners of the EROEI art are aware that issues arise in defining the "capital" factor, in accounting for differences in the modes of distribution of energy generated, and the efficiencies thereof, in projecting the benefits of additional energy supplies in societies at differing levels of energy use and human welfare, in the efficiency of capital allocation as between prospective energy generation candidates, in the apparent propensities of various societies to reinvest "capital", in the prospective capacity of societies to get more welfare out of energy use, in defining the boundaries of the "energy in" or reinvestment, process, and the like.

Probably the most difficult computations, or set of computations, to be made are those which would attempt a definition of society-wide, or region-wide, project-level EROEI necessary to sustain what humans are willing to regard as sufficient levels of human welfare ‑‑ which are informed by the great gift of fossil fuels, and what we have come to seek to maintain and achieve. Leaders in the field are aware of this shortfall in current analysis. A great deal of work, and/or ingenuity, is needed to close it.

Given the fundamental nature of the scope, stability and manageability of the energy transition necessary for all currently "developed" societies, and those which seek to be "developed", we probably need to broaden the base and deepen the insight of research and projection efforts.

In addition, efforts to standardize approaches to various elements of EROEI calculations, or at least simplify comparison and contrast between them, should be useful.

Along the lines of broadening the base of investigations, and encouraging more standardization of approaches, this author suggests exploration of a "quantitative wiki" approach to developing scenarios for future energy related investments, in Appendix C.

Though the traditional EROEI oriented calculations can shed light on critical paths through the hoped-for human civilization energy transition, there are other ways to approach the industrial civilizational need for adequate energy supplies.

At least two other analytic approaches warrant, in this author's opinion, continuing attention. They inform the question whether humanity can get through an energy supply transition without disrupting either or both of the legacy ecosystem and a functioning artifactual energy supply structure adequate to support industrial civilization at current or greater scale.

One is a large and continuing set of researches at international (and lower) levels concerning the scope and effects of "climate change" -- heating up the Earth's atmosphere. This has been led by the International Panel on Climate Change (2007). The IPCC panel has also addressed the levels of energy supply technically and economically available to human civilization. However, given the excellent quality of the work and the high level of public attention to these continuing efforts, I see no need to add detailed commentary in this manuscript.

In addition, Tim J. Garrett, a professor of Atmospheric Sciences at the University of Utah, has developed a conceptually powerful and elegant alternative approach to modeling the energy positioning and trajectory of human civilization.

Garrett works off a physical model of civilization-scale energy generation and use. He posits a set of energy potentials which flow "downhill" into the human civilization structure, through a numerable set of energy acquisition/transformation mechanisms, or systems. The human system (and the life system of which it is a part), deploys these energy flows both to build mass and structure, and to acquire more energy.

Garrett's conceptual approach accords with the Morowitz and Smith view of life's origin and, so to speak, function, of reducing differentials. That is, life is considered to be an ordering phenomenon which reduces differentials (heat, concentration, etc.).

The picture of life channeling energy back into the energy tapping processes also accords with Lenton and Watson's depiction of the life process as enlarging its energy intake, scope and mass over Earth's geologic evolution.

Garrett's construct also accords with my suggested cyclic view of energy extraction from primary sources and reinvestment in the facilities which mediate that energy extraction, set out earlier in this chapter.

Garrett chooses key elements in this process on a global scale, over decades and centuries. Among these are the scope and magnitude of the basic energy differentials available to humanity, the efficiency with which those differentials are translated into energy flow to civilization, the accumulated magnitude of the civilization structure, the historical rate of gain in energy acquisition and processing, the close linkage over time between the energy supply rate and measures of gross world product, and the relationship between the rate of production gain in the society and the rates of energy dissipation of the system inherent in its organization, and possibly resultant from global atmosphere warming.

Garrett's analysis makes unnecessary -- for this level of analysis -- many of the detailed elements of IPCC analyses and projections. Also, he does not calculate the EROEI of the various types of energy "production" facilities, or deal with the particulars of what elements of human activity should be allocated to project-level, or global, EROEI calculations. Garrett instead depicts the civilization construct as a whole as participating in amplifying the interface between civilization and the sources of primary energy. This approach in effect includes all the components of civilization -- "labor", "capital", etc.

Garrett's model includes variables to reflect the size of the energy differentials available to human civilization. He also incorporates terms to reflect the efficiency with which available energy potentials are translated into energy available to and moving through the human civilization construct.

Thus the model does not exclude consideration of efficiency or inefficiency in the process of accessing potential energy flows. Nor is it inconsistent with further, project oriented, let us say, attempts to isolate particular energy accessing technologies, and the relative efficiencies in utilization of such technologies, given identifiable inputs and energy rewards.

Among the most significant results of Garrett's modeling is identifying a momentum in a civilization, and in its tapping of sources of energy. (An advanced civilization has a set of energy production and use systems, including the mines, transport systems, distribution systems, etc., and the flows of energy through them. All this, going forward, can be said to have a momentum.)

This momentum entails and reflects a close linkage between the energy input into the system and the civilization system's feedback into the continuing process of maintaining and extending the system interface with the basic sustaining energy differentials (e.g. fossil fuels, sunlight, wind, nuclear fission, etc.)

In Garrett's analysis, gains in gaining efficiency in energy "production" and over all civilizational efficiency in energy use have had and will have the effect of enhancing the rate of civilization growth and energy flows, rather than diminishing energy flows into and through the civilization system.

Since Garrett has demonstrated, both in mathematical formalism and empirically, the very close linkage between energy flows and the accumulated gross world product over all nations and all of recent millenia, and close links between greenhouse gas emissions and civilizational energy flows, he is led to project high continuing gross world product growth over future decades, and much higher greenhouse gas levels, at any given level of energy throughput, than IPCC analyses had projected at the time of Garrett's 2011-2012 publications.

In asserting that the energy dynamics of civilization are likely to defeat attempts to solve the greenhouse gas problem by means of efficiencies in energy acquisition and processing, Garrett's work in effect directs attention to two variables involved in possible adverse effects of greenhouse gases on future civilizational function. These are (1) whether "decarbonization" of our artifactual energy entrainment systems can proceed at a rate equaling or exceeding the projected rate of growth (and prospective proportional energy intake) of our civilization, and (2) the degree of resilience, or deterioration, of the lifesystem and the civilization system to such increases in greenhouse gases, and atmospheric heating, which would occur given the quantities of greenhouse gases which would be implied in energy entrainment of the scale projected.

A third variable affecting growth of our civilization would be the scale of the energy differential available to the energy entrainment systems of civilization. (That is, are solar, wind, nuclear energy, etc. differentials of the same scale as fossil fuel differentials?) Also, as noted, Garrett recognizes the utility of gauging the efficiency of energy entrainment in his global form of analysis.

At the time of Garrett's 2011-2012 papers, the analytic community had very little information on the prospective rate of decarbonization of energy supply -- the rate of substitution of "renewable" energy production for fossil fuel energy production. Nor did the analytic community have much quantitative data on the resilience, or lack thereof, of the earthlife and civilization processes to increases in greenhouse gas levels to four or more times preindustrial levels.

Absent such data, Garrett modelled two scenarios. One was the no decarbonization -- fossil fuels full speed ahead -- scenario. The second assumed, for order of magnitude modeling purposes, a 50% decarbonization ratio obtained over 50 years. For a data modeling analysis, that was a bold projection at that time.

The first, full speed ahead, scenario depicted levels of greenhouse gas emissions over the next century over three times current levels. On some projections this could lead to over 10 feet of sea level rise. The projection also depicted very high levels of economic growth, but very likely collapsed by the drains on the human civilization system of adverse effects in the supporting global ecosystem, and also within the human civilization system itself.

Even given the 50%/five decade decarbonization assumptions, Garrett's model projected global greenhouse gas levels of 650-700 parts per million in the latter decades of this century, or about twice current levels and well above the 450 ppm level which climate change scientists might deem safe.

Where do we stand now, five years or so after Garrett's carefully reasoned and globally framed analysis?

The rate of "decarbonization", as a proportion of energy supply -- rather than total quantity of energy supply -- roughly corresponds to the rates of investment in "renewable" energy acquisition systems relative to fossil fuel systems.

In the short period of five years since publication of these papers, the decreases in the costs of wind and solar electricity generation, in addition to some subsidies, seem to have led to new investment in "renewable" electricity generation exceeding investment in fossil fuel electricity generation, according to a recent Renewable Energy Policy Network for the 21st Century (2016) report.

The magnitude and direction of the "renewable" energy cost reductions constitute a vector pointing toward substantial and rapid, but not complete, decarbonization in the electricity supply sector. Bloomberg's New Energy Finance group (2016) currently projects that by 2040 -- less than three decades from the 2011 papers -- zero emission sources will account for 60% of worldwide electricity generation capacity. If realized, this would reflect a rapid decarbonization rate.

However, substantial investments in complementary facilities (forms of energy storage, transmission, and other supply/consumption matching systems) are also required. And electricity currently accounts for only about 20% of global energy consumption.

There is as yet no visible evidence of rapid decarbonization in transport and heating functions.

There are, however, visible potentials for these functions. The 2008 Sandia report previously mentioned suggests that at about 1-2 cents per kilowatt hour, there could be a potential for mass production of chemical fuels. Some of these could result in decarbonization.

To the considerable surprise of some, including this observer, there are currently some reports on wind energy sales in the 2-3c kwh range (though this reflects significant subsidy -- perhaps in the 2c kwh range)[55]. Credible projections suggest further cost reductions of 35-40% by 2050. Wiser, et al (2016).

As of this writing the trajectory of solar costs nears the 5-6 c kwh range.

Just where the relative costs of electricity-to-chemical fuels and fossil hydrocarbon fuels, fully accounted for, will be 2-5 decades out remains to be seen, and may vary considerably in various locales, but narrowing of the differentials seems to be under way.

Also, in parallel, taking into account these decreasing solar and wind power costs, and projections of similar cost decreases in battery costs, the Bloomberg group projects that 35% of light duty vehicle sales will be in electric vehicles by 2040.

---

55 See for example a 2016 Department of Energy report available at http://energy.gov/sites/prod/files/2016/08/f33/2015-Wind-Technologies-Market-Report-08162016.pdf.

This would imply a substantial decarbonization rate in the propulsion of vehicles, assuming there is substantial decarbonization of electricity production.

There are potentials for the use of solar energy for heating purposes, and increasing the use of nuclear power for both electricity and heating purposes. However, the rate at which these potentials will be realized is not clear to this observer at this time.

Where do we now stand in referring to Garrett's proposed landmark, the question whether humanity can achieve a 50% decarbonization rate for the entire range of its energy generation facilities, in five decades? One can see evidence of some significant decarbonization of global energy supply. There is a vector of some significant scale pointed in the right direction. But in the opinion of this observer the question is not yet answered in a clear affirmative.

Garrett does not engage in detailed discussion of the resilience, or lack thereof, of the lifesystem, and the human part, to the higher levels of greenhouse gases which historic and current data suggest are likely in coming decades. His projections do suggest that the energy demands inherent in the size of our civilization, and its growth trajectory, will produce atmospheric carbon dioxide levels which are likely to have substantial economic and ecological effects given the rates of "decarbonization" which could be documented or confidently projected at the time of his publications.

This is all a work in process. The inquisitive and concerned reader, and researcher, will have no lack of data tables and projections with which to track the rates of production of various forms of energy and of decarbonization in energy production processes in coming years. Data and projections are driven at university, national, and international agency levels, as well as various interested "private sector" or non-governmental organizations. And should be. The evaluation of such data and processes is an ongoing requirement in our busy efforts to keep our civilizational machinery intact and on some sort of sustainable track.

I highlight Garrett's recent work because it relates the human civilization system to underlying physical processes, at a usefully conceived basic level, puts civilization energy entrainment in a framework consistent with the nature of the life process, and has interesting and potentially useful mathematical arguments, as well as taking into account the propensity of the life process to reinvest any gains above bare subsistence into more energy acquisition and life expansion.[56]

As to the momentum of our civilizational growth, the recent worldwide declines in productivity growth are an important signal needing careful attention

---

[56] Garrett (2012) was sufficiently candid to assert that if his judgment that there is an invariant link between energy flow rate and over-all size of the civilization construct is not borne out, his form of analysis loses value. (p.5) But in my view the form of analysis itself contains substantial value, worthy of examination, and elaboration, or refinement, by other researchers and modelers. Garrett has in recent years extended his form of analysis to project the long term growth projection of human civilization, with noteworthy results. Garrett. T.J. (2015)

and interpretation, as the foregoing treatment of EROEI issues in this manuscript suggests.[57]

Garrett's form of analysis, and well as the analyses of a great many other researchers, are in accord with the general impulse of many academic and policy leaders to curb fossil fuel use and foster renewable, or sustainable, energy production as fast as may be feasible. In the preponderance of academic, policy, and public opinion, the needed direction is increasingly clear. The prospects of success in avoiding overshooting our ecological base would appear, at this point, less clear.

Humanity has a two-century stress test coming up. This chapter seeks to frame this stress test in the context of the grand themes of order development in our universe, and on our planet, where Earthlife has barely touched the energy fluxes around it. But those grand themes frame an existential challenge which we have every reason to take very personally.

For the moment, let us make the critical, and rather heroic, assumption that humanity may be successful in maintaining and increasing artifactual energy supplies at levels exceeding those of the early part of the 21st century, CE, for the indefinite future.

We would then continue to be faced with issues as to how to integrate the additional artifactual energy flow and the legacy photosynthetic energy system which sustains most of Earthlife. Humankind would be altering Earthlife from a single major energy supply system to, or toward, a permanent (as long as we are around and well organized) dual energy supply system.

This raises the prospect of "crowding out" some of the legacy system, and, more generally, the challenge of managing additive integration of the two energy streams, as distinguished from destructive interference between them.

The "problem" is made even more challenging because both energy supply systems are and will be complex, we have only begun to understand the interdependence dynamics of the legacy system, we do not completely understand the energy system we have been building, we find it difficult to predict the results of our actions, and we have major problems and challenges in composing and coordinating the organizational systems and ambitions of our recently-globalized human social complex.

In sum, in legacy Earthlife, the energy dynamics formed a sort of hierarchy off one major base. The photosynthesizes captured the bulk of the energy, the grazers harvested a portion of that, and the predators harvested the grazers, up to the peak predator. This is oversimplified -- there is a food web -- but almost everything on the surface of the Earth worked off the photosynthesis base. With agriculture, we humans

---

57 I earlier note the conceptual possibility of EROEI levels which would not sustain substantial productivity growth. Garrett also notes the possibility of stasis points in the evolution of civilization potentials, from a different analytic perspective. Whether current phenomena signal any slowdown in the energy intake and reinvestment levels of civilization is of course to be played out in the 'real world'.

began a tendency to sip from almost every cup in the ecosystem, but we were still working off the photosynthesis base.

When we harnessed wind and water power in our economic systems we spread our net wider. When we started to exploit fossil fuels on a large scale, we made a major shift to powering what we call our economic systems from a base other than current photosynthesis.

Now our transport systems, our manufacturing and assembly lines, our cities, and our communications all work off energy flows outside the legacy current-photosynthesis energy flows. We call this the Industrial Revolution. Our economic and social systems – as systems – now have a metabolism not based on current photosynthesis.[58]

Now that we plan to continue dependence on artifactual energy sources --solar energy, nuclear power, wind, geothermal, etc. -- for our "economy", we continue to move toward largely bypassing the photosynthesis system except for fueling our bodies. This currently does present a number of problems, and would continue to do so.

In the last chapter, I will address an orientation toward these problems. In anticipation, the thrust of the thought is that we should seek to make these two energy stream complementary, in service to a larger and more complex earthlife system, rather than taking an entirely human-centric approach.

It is our social organization which has brought us to this need, and to this prospect. It is our social organization which, in interaction with the universe's characteristics, will determine whether we develop more of life's potentials for ordered states, or do not. Subsequent chapters addressed to "globalization" and "ethics" will address what it may take to measure up.

---

58 If and as such a social system were to reproduce itself, it would have to reproduce its, or at least an, energy supply system as well. Taking the hypothetical Moon-bud example, the energy supply system might be solar energy.

10 Globalization (Make It or Break It)

History is not ended, in Fukiyama's phrase. It is in a new-construction stage.

In the language of the section of "emergence" and hierarchy, humanity has been in the compositional stage of building the next hierarchical level in human organization -- global integration.

In doing so we are in accord with the nature of thermodynamic and order building systems. In other words, the way the universe works projected in this document would seem to make this next step of global organization potentially available. This can be seen, in the American idiom, as "manifest destiny" – if we can pull it off and sustain it.

But the Universe cranks out its constructs in a probabilistic process. The thermodynamic forces might be with us, and the probabilities might favor us. But there is no guarantee of which I am aware that our particular type of animal will have what it takes to achieve and sustain a high level of global social functioning at this time on this Earth.

As of this writing, a set of circumstances has raised questions about whether tendencies to erect global trade patterns, travel, information flows, agreements, and institutional structures will slow down, pause, or even reverse. Countries in the European Union have substantial "Eurosceptic" political parties, Britain has seceded from the EU, and a backlash against new trade agreements has appeared in the United States.

This has occurred in rough simultaneity with a major slowdown in per-person productivity gains globally, perhaps most notably in the Euro-American complex, for over a decade, widespread concerns that the "middle class" in industrialized countries have had slow to no income gains going into the second decade of the 21st Century, and is getting "squeezed" by gains in higher education/income cohorts, and scholarly advice, such as that by Robert Gordon (2016), that our society seems not to be generating major welfare-raising innovations at the rate it did in the 20th Century.

However, even if human civilization does not make the great energy transition discussed here, and retrogresses -- evening out, let us hypothetically suppose, at about the High Agricultural stage -- global interactions of some sorts seem likely to exist. After all, Europeans claimed the American land masses before industrialization took hold in the 19th and 20th centuries C.E.

This monograph supposes that momentum toward high energy civilization persists, and there is reason to try to anticipate the requirements of a more globally integrated and globally conscious human civilization.

We do not know as yet just what forms and degrees of global integration will eventuate, and whether forms and degrees will rise and fall, build, collapse, rebuild, etc. One could dash off many possible scenarios. We can predict, I suggest, that the extent to which humans are able to construct an effectively functioning global social system will have much to do with how we will live and what we will amount to on this Earth.

Briefly to review, why would we suppose that we humans might realize a globally functional and productive social system?

Ants and termites explored large-number socialization earlier than we have. They have global effect. But they do not have global hives. Why should we have peacefully, globally networked human hives?

We are a much larger animal, mobile over a much larger range, with substantially more computation capacity per each, and with communication systems which embody substantially more capacity both to image the universe and to organize concerted activity.

This has been a long time coming. The homo sapiens which left Africa @100,000 years ago took a large first step toward globalization, simply by reaching all habitable areas of the globe and simultaneously eliminating their competitors. But their social structures were localized, not global.

With agriculture, starting roughly 10,000 years ago, began a long period of bloody attempts by organization centers, marshalling tens of thousands to millions of participants, to extend their control as far as they could go. Some of these attempts gained control of areas stretching for thousands of miles, for time periods ranging from decades to centuries.

In the last few hundred years, Europe launched some forms of globalization –i.e. global reach of Europe-based control systems. Europe used "guns, germs, and steel" in Jared Diamond's phrase, plus, importantly, boats.

England was an highly successful aggregator, or colonizer, of this era. England did not depend as much on massive destruction of competitor organization centers and, to an extent, their populations, as did many prior social organization centers – like for example Alexander the Great and Genghis Khan. England used a combination of force, installation of governors, co-option of local control centers, transplantation of colonists, and goods exchange systems. It overran where it could, as in Australia and the United States, coopted where it could, and used force of arms selectively and strategically. England apparently learned much from prior empire builders like the Greeks and Romans.

But the end of the "second world war" in the middle part of the twentieth century, was followed by the end of the European global colonial empires.

For the last half century or so, humankind has been attempting to build consensual global coordination systems. These include a host of bilateral and multilateral agreements among "states"; and a rudimentary proto-governance system called the United Nations. The UN has at present very limited resources and authorities, and is loosely patterned on the representative congresses of England and the United States. There are smaller coordination groups such as the "Group of Eight" major economies, and a group of 20. There are specialized bodies such as the International Court of Criminal Justice and the World Trade Organization. And there is considerable scope for "non-governmental" organizations, locally and globally. Behind all this are multinational organizations, and international markets for "capital" (i.e claims on resources), goods and services.

So here we are. Looking forward from here, instead of joining in the well populated field of imaginative scenarios for our future, I propose to identify some of the priorities, and constraints, suggested by the foregoing discussion of the characteristics of how order is built in the universe.[59]

10.1 Keep the Energy and Materials Flowing

This is a prescription both of what we should be able to see as both "common sense" and a result of organization theory. In Chaisson's terms, the 'free energy' in the global system supports the energy density and complexity of the system. The resources which support large energy flow are distributed asymmetrically world wide. Sophisticated exchange and combination (e.g. fabrication) facilities are needed to maintain and develop them. If global flows of energy and material break down, so does globalization.

If we do not successfully meet the fundamental challenge of transitioning from fossil fuels to other large scale energy sources, a global civilization will be poorer, and might amount to a thin and easily broken veneer. Making this transition, however, may put huge, and possibly unsustainable, stresses on any hoped-for global polity.

My inbuilt, virtually instinctive, response to such a situation is to attack the problem. That is, globalization should be constructed consciously and determinedly to help maintain high global energy levels in the human population and to make successful the energy replacement transition from fossil fuels.

However, we also face the task of fitting all this human-centered energy flux into a workable, and hopefully augmented, life system on Mother Earth. More about that later.

---

59 The rationale for a globalized economic and social system which I offer is not exactly the same as that of current economic theory. I do not suggest that the ideas of gains of specialization and trade going back at least to Adam Smith are wrong. They are included as subsets in an argument having to do with mobilization of energy flows, the advantages of large scale combinatorics, and energy minimization bases for "specialization". Thus there are similarities in the suggestions made here and suggestions based on conventional trade theories.

10.2 Maintain Systems Which Facilitate Realization of Combinatorial Potentials

We now gain from a 'liberal' international trading regime which mediates complex assemblages of foods, materials, techniques, and the like on a global basis. The 'market' institution is one such mechanism. A system of autarchic, contending states with hard borders, frequent conflicts, and little commerce between them would constrain the realization of combinatorial potentials. And put us all at great risk of losing the game entirely.[60]

10.3 Recognize that Coordination Entails Constraint

The foregoing chapters on the correlational basis for order development, and hierarchies, make clear that it is, always and necessarily, constraint which gives rise to the next level of organization. That constraint creates ordering potentials may seem a paradox in the building of order in the universe. But it is inbuilt in the way the Universe works.

Let us take counsel from our own bodies. The intricate treaties between cells and cell lineages in biological systems were felt out over hundreds of millions to billions of years, on an unplanned, trial and error basis. Now, in multicellular organisms, only cancer cells range freely, and in doing so they destroy their hosts and ultimately themselves.

In our improvisational 'cultural evolution', we humans have been engaged in erecting tiers of human organizations over the last ten millennia. Experience, trial and error, and foresight allowed by our evolved collective representation systems (language and derivatives, and graphical systems) have all played roles in this evolution. But each next stage of group building obviously does not always go easily.

Large and powerful organizations, like the United States, my own venue, may be particularly reluctant to cede autonomy. Indeed, we balk frequently, and tend to attempt to use international coordination systems in an US-centered way.

We are not alone in this tendency. But we do need to 'get over it', in the sense of knowing that some such cessions are necessary.

This does not mean that we do not consider carefully what to cede, or delegate, how much, and under what conditions. As before noted, the intricate constructions of life units has proceeded through hundreds of millions of years of trial and error. We literally embody much experience in group construction, even though most of it has been

---

60 This means maintaining robust trade networks within and among nations. As a subset, the United Nations Environmental Program has taken an initiative to stimulate trade links among developing countries as to use markets to increase sustainable development.

unconscious. At every level, the pull and tug between self preservation and autonomy on the one hand and collective constraint and commitment has had to be negotiated.

And so, nationally and internationally organizational units which are the building blocks of global integration need carefully to define how best to yield to constraints for a greater common good, while preserving necessary component integrity, and productive degrees of freedom, within the international systems to be constructed. Very likely this will all be done on an incremental basis. But this is our task.

10.4 Guard Against Parasitism in Our Social Systems at All Levels

Again, we can take counsel of our own bodies and those of other multicellular organisms.

Leo Buss, a biologist with a penetrating grasp of life organization and evolutionary history, pointed out that in multicellular organisms cell lineages could, without constraint, continue proliferation well beyond the point at which they served the organism as a whole.

In complex human organizations, we recognize parasitic drains on the social system as 'corruption' and 'free riding'. In our ten millennia or so of social evolution, we have explored this unfortunate theme repeatedly. We see it played out in several situations in our 21st Century AD. Ruling elites can easily -- indeed, eagerly and stubbornly -- become parasitic.

Restraining parasitism is particularly important in any globalization resulting in one central coordinating system. In biological organisms, competition between individuals and groups in fitting into the world tends to weed out the parasitized individuals and groups. But a global human society would have no known competitors at its global level of aggregation. That could make for a very sick and stagnant structure, absent well designed and well maintained systems to eliminate parasitism.

One of the major advantages of the democratic form of social organization is the controls on parasitism by those provisionally granted access to and disposal of community resources.

10.5 Apply the Tools of Relational System Analysis, Including Social Network Analysis, to Issues of National and Global Political and Economic Development

I have pointed out the work of Barabasi, Ulanowicz and others in relating life system and ecosystems to basic tools of information theory. We see the beginning of efforts to re-cast the economics discipline in these terms. In recent decades, the explosion of 'social network' theory tells us that a globalized human society is and will be a global relational system, and a global system of social networks.

As we pursue this sort of analysis, we will need to pay close attention to network integrity issues. For decades, researchers such as C.S. Hollings and Robert Ulanowicz have pointed out that complex ecosystems could get into 'brittle' states, or states too near the 'boundaries of (their) window(s) of vitality'. (Ulanowicz, 1997, p. 156). Joseph Tainter has made similar observations as to complex human societies. The global Great Depression of the 1930's and the financial shocks of 2008-2009 have provided real-time examples of how a complex system can undergo an abrupt and painful reorganization to lower levels of activity. We will need to be paying close attention to how to achieve useful, but not stifling, levels of homeostasis, and as we do so how to avoid dangerous 'brittleness' in globally interconnected systems.

10.6 Recognize that the Power Law Rules

This is a necessary outgrowth of using the tools of relational system analysis. But this recognition is not easy for us humans to swallow. We are constantly trying to assure that participants in our democratic social systems have 'fair' access to the opportunities it affords, and at least a minimal level of sustenance. The tendency to try to make sure the neighbor does not 'cheat' to get advantage is built deeply into our primate software. That some of us will make out much better than others is hard to face up to, even though it happens every day in every area of human endeavor.

However we may feel about it, the 'power law' distribution is built into the fabric of order construction in the universe. Economic measures over time show this distribution of wealth in every polity, and every era measured. It seems about as ubiquitous as the second law of thermodynamics.

This means that an utopia where everyone is completely equal in circumstance -- and by the way every child is above average -- will not work.

We can reconcile this with a powerful and perceptive prescription of John Rawls -- that we should be willing to subscribe to a social system if it serves group welfare, as if we do not know where we would fit into the system. We can also require rules of the social game which forbid lying, cheating, and stealing, so that both individual and common interests are served. We must limit parasitism, whether in the name of the power law or otherwise. We can outlaw predation, in its many forms from dishonest bargains to genocide. We can try to give everyone who plays by productive rules opportunities in the game. We can try to have systems which afford all participants an opportunity to be rewarded according to contribution to the society's collective welfare, But the power law rules. We can't make each and all come out equal.

10.7 Maintain Fidelity in and Extend the Reach of Our Collective Representation Systems (Language, Language Derivatives, and Graphical Systems)

Giving this objective high priority in constructing a global organizational level may seem a bit odd. Prioritize integrity in our representation systems? Just what does that mean, one might ask.

Language and associated systems provide Homo Sapiens a large part of the advantages of speed and scope of 'cultural evolution' as compared with genetic evolution.

This collective representation system is so important to us that we spend large amounts of our lives learning it and its contents, maintain large and expensive institutions to serve as repositories and promulgators of it, and expend considerable resources systematically extending it.

With these representation systems, we can encode experience, or history, both individual and collective. We can encode current action plans, on the fly. We can encode organization schemes; project futures, effect coordination among ourselves in both simple and complex matters. We need to do all these things to construct the next levels of organization effectively.

But, of course, our representation systems have often been used for parochial purposes at war with the general group interest. We have seen societies create myths to maintain order within themselves (whether such degrees and types of order are optimal or not) and to motivate conflicts abroad. We have seen integrous explorers of universal mechanics, such as Galileo, suppressed to prevent what social organizers considered might lead to questioning of the social order extant in his time and place, and of their role in that social order. We have seen fraud and destructive opportunism facilitated by language at all scales of human endeavor. I think I need not catalogue the full scope of abuses of representation systems – you, the reader, can seek such catalogues in many places.

Though "learning" has been prized in many societies, only in the last few centuries, and largely in the West, have we evolved the rigorous codes for inquiry and reportage which we characterize as science, and also institutionalized the "right" to the free and open public communication concerning our political and social institutions and actors (which we characterize as freedom of speech). Both institutions have enormously benefited those societies which have used them consistently and thoroughly.

Science is now prized around the globe. However, free critique of the organization of communities, which we call political organization, is restricted in many large polities.

I would argue that these institutions of disciplined, integrous inquiry and communication will be of great value to all participants in any attempt to construct an efficient and sustained global civilization.

If a polity cannot force on the remainder of the world community the institutions[61], or political arrangements which it favors, it needs at least to be able to trust the institutions to which it must agree and on which it must rely. As Deudney and Ikenberry (2009) have pointed out, this applies to both democratic and autarkic polities.

Fortunately, there is reason to think that many or most of the polities participating in constructing a globalized society, in a consensual way, may come to recognize the need for these institutions in the international arena (even if some or many of the ruling elites at this time would prefer not to see them in their own societies).

Let us take "science" as compared with "statecraft". A great deal of craft has gone into statecraft over the centuries. As an analogue, a great deal of highly impressive craft obviously went into the Hagia Sophia, beautiful Moslem mosques and Gothic cathedrals -- all before we had materials science and formal engineering disciplines.

But we babbling apes are going to need every bit of scientific understanding we can afford to obtain if we are to build effective, efficient, and lasting global institutions. We need to draw on all the sciences to construct a verifiable and effective science of human organization. We are going to need insights into our own evolutionary history, such as are explored by scholars and other participants in the Human Behavior and Evolution Society. We are going to need sound, advanced economics understanding, including economic analyses taking into account non-equilibrium thermodynamics. We are going to need social network analysts. We are going to need voting systems analysts.

We are, I suggest, going to need an international community of scholars, analysts and commentators whose interests are not tied to a particular ruling elite, or other specific economic interest, dedicated to searching out verifiable, reliable understandings of how social systems do and can work – not only in broad information theoretic terms but in the details of how those terms translate into human institutional affairs. And our Departments of State and equivalent should have close ties to and take instruction from such a community.

This is the "science" part. Now to the associated "free speech" part. Participants in the global institutions will be served, I suggest, by "transparent" arrangements which allow reliable monitoring, and by ongoing, active, widely based, unsuppressed critique of the institutions and their actors. Vigilance – active, informed, outspoken vigilance – will be the price of liberty and good order globally, as it has been in our more fortunate local and regional societies.

---

61 We might stipulate that almost any society might prefer, if it had the option, to conquer the world and rule the world as it pleased in its own interests. But as the world is composed today, and is likely to be composed in the foreseeable future, no one national society, or coalition of societies, would seem likely to be able to organize the world by force of arms, and then maintain any order constructed thereby. Very likely, any attempt by a large polity to organize the world by force would lead to results dwarfing the destruction of World War II.

Were it not for control of parasitism, one might limit this understanding of "statecraft" or "state science" to the functioning coordinators, or rulers, at each level in the human hierarchical structure. But parasitism is endemic at all levels. The dispersion of this function to other elements in the society would seem necessary to provide effective check on rulers, and would also seem likely to provide more depth and diversity of understandings.

10.8 Build Regularity Into the Compositional Arrangements of a Global System. (This Can Also be Characterized as the "Rule of Law")

We are starting with a very heterogeneous collection of elements on the international scene. At this point "states" come in a wide variety of sizes, resources, degrees of organization, and wealth. And we have a very wide range of coordinating mechanisms.

However, I have pointed out that regularities – reduced variance, or constraints in variance – are what create differentiated relational regimes -- which build the next level of organization in hierarchies. In human terms, we have to have regular, dependable arrangements between elements in each level of society if we are to have a functioning society at that level.

This means that there must be an effective "rule of law" (whether it is called that or not) internationally as well as – we would hope -- in local and regional states, notwithstanding those who deny the existence of and need for a regularized, institutionalized international order.[62] See point 2, above.

The founders of the American republic appear to have been aware of the internal wars over succession to power, erratic taxing policies, and imperial whims which plagued the Roman Empire. Thus the regularity of our transfers of "power", or coordinating authorities, in democracies, and our emphasis on the "rule of law" in our internal affairs. We face the same considerations and requirements in erecting the next stage of social integration.

---

62 I project the role of "law" as paralleling the role of genes. That is, codes can specify the procedures of organization building and organization function. But those codes respond to and correspond with thermodynamic imperatives. This view is in some respects analogous to "natural law" concepts of earlier centuries. It puts the nature of nature first, and law as the tag-along handmaiden. And it offers a concept of the regularities of "nature" to which law may be anchored.

At present the conventional view is that there are three types of international law: (1) "public" international law – the regularities of relationships between states; (2) the laws relating to supranational institutions (e.g. the European Union); and (3) "private" international law – the laws governing activities of "private" actors operating internationally. These may be seen as craft categories. Law is now taught and practiced as a craft. At a high conceptual level, I here suggest, law in all these categories needs to be tightly tied to evolving concepts in social organization.

A substantial question is whether the global integration level should get involved in the internal activities of its components. We face this issue now in international affairs.

A basic characteristic of correlation processes is that the next level up "sees" as its components, so to speak, basically the sub-component as a unit, rather than the actions of the individual elements within the component.

But we have seen that life units have capabilities to organize their own internal states so as to affect their components, reaching "down", and also to affect the next level "up" of organization. We cut out tumors in our bodies and we draft our community laws. In commerce, franchisors dictate in detail the organization of franchisees. At the international level, the World Trade Organization and the European politico/economic organization set standards for admission to their collectives.

"Failed states" result in ineffective mobilization of resources which could in well organized states contribute to international trade and wellbeing, and failed states may also loose toxic elements upon other elements (such as trade ships) near them.

So we may expect some degree of "top down causation" in a globalized social system.

This presents both perils and possibilities.

We fear stultifying and unnecessary uniformity on the one hand, and on the other hand the incubation of rogue actors on the international stage in poorly organized and remote venues, such as the Al Quaeda and ISIL organizations.

Americans have had several experiences in recent decades demonstrating how difficult is reorganizing other States, within the international community. We imagine and hope for the resolution of local conflicts – state and sub-state – curbs on weapons of mass destruction, the recognition of "human rights" which enable productive and peaceful function within member States, etc. But we are not demonstrably highly effective at bringing about such things, on any sweeping basis, and even in particular circumstances, at this time.

10.9 Create Coordinated Control of the Emergent Effects Of an Integrated Global Human Society

First let us focus a bit on the unique and difficult situation facing consciously coordinated life units in an organization of life units, when it comes to the "emergent effects" of our organization. Water molecules in a falling raindrop do not and can not gauge the effects of the raindrop on a human cheek, or where they will individually be when the drop splats on a rock. Nor can those molecules do anything about either situation.

The ants and termites in their hives do not attempt to image the hives, or their effects on the ecosystem into which their hive fits. Their patterns of interaction between themselves and between their hive and their environments are negotiated in the evolutionary give and take described before.

And so it may be with us, if we do not and cannot manage ourselves. But we have hopes for our "cultural evolution" potentials.

If we humans are to use "cultural evolution" to shape our own social organization, and to manage our relationship with our environment, we individual human mites must collectively image our human whole, and image its setting, and then agree on concerted action within the whole. When our society simultaneously builds upon and affects its life setting, we must image and try to manage, together, how we sustainably shape both selves and environment.

We, life units, are the universe updwellingly emergent into recursionary self management. That's a mouthful. It is also a handful. Pulling that off is going to be a neat trick.

Now back to what we can consider our "emergent" effects. In previous chapters I have urged good bookkeeping in distinguishing between the construction, or composition, of a coordinated relational regime, on the one hand, and on the other hand the relationships of that regime on aggregates external to it. The latter, I suggest, is what we may characterize as "emergent" phenomena.

Those concerned with "sustainability", the ecosphere in general and species loss in particular, and global warming, have in effect been pointing with considerable and increasing concern to "emergent" effects of the expanding human hive. Those effects include diminishing global forests, increasing species extinctions, diminishing ocean life, and a warming global atmosphere.

Many are concerned about these emergent effects of our hive activities primarily because they may undercut the support for our hive. (We see everything as support services for us.) In a subsequent section I will address a larger and I suggest more useful perspective. For now, this is at least a beginning of some wisdom.

The insects and the other animals in Eden have not been endowed with the capacity for choices as to their "emergent" effects on Earthlife. We have been. Best we use such capacity as we have.

As it turns out, we humans may choose not to manage our emergent effects. Or we might take some actions evidencing some intent to do so, but shrink from taking those actions required for us to do so. Or conceivably we might devise institutions which do effectively control the emergent effects on earthlife of our human activities.

The fate of billions of our descendants ride on our choices. As for Earthlife itself, whether its human socialization experiment will be a blight upon it or a boon for it remains to be determined. At this point, we have eliminated significant portions of legacy Earthlife. What will replace it is to be determined.

We can see that we are doing pretty well in organizing systems to monitor, at least, our human-hive effects on various elements of our environment, at local, state, national and international levels.

We tend to track most closely the "mutual information" between humanity and the earth environment – that is, relationships which most matter to humans between earth and earthlife on the one hand and humans on the other hand.

Examples of such measures are land per human (or collection of humans), water per human or collection, air quality to humans, viewscapes available to humans, animal products per human or collection, and the like. This "resource" access or availability for humans is mixed in with descriptions of state of the life system as a whole.

The implicit perspective is that the Earth as a whole is humanity's life-niche. One might derive from this a view of Earth as Earthpark, and humanity as the park rangers. A perspective on this from a passionate lover of wildlife, in the footnote, captures what we are apt to preserve as "wild" in our current trajectory.[63]

For ongoing sources of information of the effects of humankind on the earth and actions to moderate or change those effects, one can consult national environmental protection agencies and "private" non-profit environment protection organizations, such as the World Wildlife Fund, the Nature Conservancy, Friends of the Earth, etc. (and their analogues in other venues). However, I particularly recommend the United Nations Environmental Program, since it is globally positioned and at this point competently run.

As to effective systems to control our collective impact on Earth and Earthlife, as distinguished from monitoring and reporting systems, the picture is not as good.

In the United States and Europe, we have extensive control mechanisms in place designed to prevent effects on the "environment" which directly affect human welfare – water quality, air quality in cities, ingestible chemicals, etc. We also appear to have had some significant effect in moderating (but not eliminating) large animal species loss, and in maintaining some landscapes in something like their "natural" state. We have preserved at least some pockets of legacy nature, and not gotten around to disturbing some other areas.

As of this writing, UNEP report "GEO5" identifies as major, unchecked or inadequately checked effects of humanity on the Earth system the warming of the planet's surface (from greenhouse gases), the reduction of fish stocks in the ocean and severe deterioration of the biome in some coastal ocean areas, species loss over a broad range, tropical forest loss, and enlarging the atmospheric ozone hole.

---

63 "The question is not whether we must manage wild nature, but rather how shall we manage it ... There is no pristine nature to conserve. Only those unaware of the past can imagine that any ecosystem is unaffected by humanity.... All conserved wildlands are ecological islands in an agroscape, an urbanscape, and will be so forever. Even if they are permitted to survive as self sustaining wild lumps, which is what we lobby for, these lumps will melt, shrink, homogenize, evolve, and be washed over, inexorably, inevitably, mercilessly.

Sustainable agriculture has been around a long time. Let's have sustainable wildlands... They need market development, crop rotation, experiment stations, subsidies, insurance, innovations, entrepreneurism – and they need to pay their bills, be a producer, be open around the clock, and be welcomed at society's table." Jantzen, Daniel (2000).

From our human perspective, we often see the problem as "the tragedy of the commons". When all draw on or affect a common "resource" (from our perspective), like the ocean and atmosphere, but no one can control the collective effect, the collective effect may be to exhaust or seriously damage the utility (to us) of the common "resource".

The UNEP "GEO 5" document also summarizes a variety of "economic" instruments which have been used at various times and places to affect human activities which affect the environment, including assignment of property rights, creation of markets, "fiscal" instruments like taxes and emissions charges, charges from controllers of common resources to users of those resources (like park access charges), financial instruments such as tax discounts and rebates, legal liability assignment systems, and financial instruments such as bonds and deposits (to raise funds for conservation projects) . Among regulatory approaches, the "cap and trade" system for greenhouse gas emissions (cap the total emissions and allow trading of emission rights) is currently favored by not a few economists.

The problem is not so much the availability of ideas about how to affect human impacts on the ecosystem. The problem most urgently facing us is devising means to get workable global concord on putting into effect some of those ideas.

Multilateral treaties between nations have multiplied in recent decades. If the tendency toward centralized coordination in human governing bodies continues – and that would seem likely to me -- we may see more delegation of coordinating functions to UN-like entities in the future.

Finally, in this section, as a sort of summation, one could posit that the recently energy-hyped global human hive system and its global environment will in the future co-evolve. On the human side of that coevolution, we are now aware that we are dealing with two distinguishable but related forms of evolution – genetic and cultural.

That coevolution might hypothetically be smooth and gradual. But it may not be. In the time scales of genetic evolution our human hive building has seemed quite abrupt – covering only the last 25 or so millennia, or 100 or so generations of humans.[64] And if the fossil fuel boom passes without large scale energy replacements, the boom of human population and some of its ecological impact will also abate, making this last and next few hundred years a blip in evolutionary time.

In this new form of evolution, human intentionality may play a significant role. If it does, for the first time in the evolution of life, evolution, and the universe, may come to steer – or at least help steer -- its own course, to shape, within the imperatives of its thermodynamic base, its own history.

---

64 The suddenness of this transition suggests, to this author, the primacy of non-equilibrium thermodynamic forces as compared to 'genes' in shaping human and other life. We appear to be in effect propelled, or hustled, into hive living, with our genes possibly lagging behind, by the dynamic imperatives of agriculture and industrialization.

Or perhaps this may not occur. The reach of our collective intelligence may be short. We may be able to see little into the future, because our universe operates probabilistically, earthlife is so complex, our impulses and appetites control us, and our intellect is so limited. We may be able to sustain collective shaping efforts only fitfully and ineffectually. How will we sustain consequential coordinated global efforts over millennia, particularly efforts which deny our immediate, compelling, parochial passions?

As we look around the globe at the emergent effects of our global human civilization thus far, many of us question whether life's evolution of our form of intelligence may, paradoxically, be a blight on earthlife as a whole, rather than a means of extending its scope.

On the other hand, if humanity can coordinate its fractious components, restrain its unruly appetites, and incentivize and shape the undirected evolutionary urges of the remainder of Earthlife, conceivably, and possibly, just, one may hope, possibly, humanity might help make a more ample Earth, in which much of our own appetites can sustainably be incorporated. That is the issue the next chapter addresses.

10.10 Maintain Sufficient Continuity in Organizational Structures to Prevent, or Limit Effectively, Massive Breakdowns in Economic and Social Systems

The objective speaks for itself. However, managing this, while also having dynamic and progressive social systems, and preventing ossified, repressive, exploitative political and social elites, will demand skill, and perpetual vigilance.

10.11 Be Patient Enough to See the Project Through

There are at least basic aspects in which endurance and persistence will be required in meeting the challenges and opportunities of the Antropocene.

The first is in the continuing requirement of productively reconciling into a sustained ecumene fractious global humanity. We are so many, so short sighted, so various, and so contentious.

Perhaps we might start by realizing that in the making of hierarchical life, we must become parochial before we can become unified. That is, in making groups of groups, each group identity must be formed, tested, and viable before it can be reconciled with others into the next-up level. We speak of bigotry, schisms, clientelism, and other deplored limitations when we have become conscious that a next level can advantageously be formed out of what we are so proud of having advantageously formed to begin with. And the composition of groups has often been experimental, intuited, and heuristic as much or more than master-planned.

I have argued for more 'science' and calculation in "statecraft".' And we do now form social institutions with some foresight and intention. But building a sturdy and productive world order is and will be a lengthy and tedious business for us human mayflies.

Secondly, as is noted above, the needed energy source transition seems likely to be decades in length, at best, and may be economically and socially difficult. Shifting gears between the enormous boon of fossil energy discovery and exploitation and building up an adequate structure of artifactual sustainable energy supplies, possibly having somewhat lower ratios of energy returned on energy invested, may involve periods of relatively low growth and disappointed expectations. We will need a global public which understands, accurately, based on truthful information, the process involved, and can have the vision and patience to maintain social discipline and cohesion. (While not falling victim to exploitation by unprincipled leaders.)

In setting out these criteria, or guidelines, above, I am well aware that I have projected as grand themes social conventions which some might deem parochial rather than needfully universal – that is, projected to a larger scale particular forms of governance currently favored by just a subset of humanity's teeming billions.

My defenses are four.

First, I have made an effort to propose considerations consistent with the theoretical underpinning.

Secondly, one goes with what one considers to be the best insights currently available.

Thirdly, I have suggested some things which democratic societies are not notoriously good at -- such as recognizing international constraints, taking good care of the life system which created and sustains us, and recognizing that we cannot democratically dry every eye and achieve perfect equality, domestically or globally, among all our brothers and sisters. I am proposing utility in doing some things democracies seem to do imperfectly, rather than assuming perfection in the current forms of democracy.

And lastly, other persons and groups from "Western" and other societies are perfectly capable of either agreeing or disagreeing with all the premises and conclusions in this book, and deriving their own inferences and conclusions for collective review and discussion.

11 Ethics, Thermodynamics, and The Anthropocene

This is the section where I suggest where humanity should be heading, on a basis broader than "globalization".

This is my thesis. We should be seeking to exploit the "non-equilibrium thermodynamic" potentials for order building on our earthly life base, helping evolve a more ample earth life system, and any feasible extensions of it. We will need to be devising, or evolving, the forms of human social system construction which will enable us to do this.

As background, all of us, in our journeys through life, have heard the questioning about so-called existential issue of the "meaning of life", and the related questions about what our human societies should be maximizing – "greatest good for the greatest number", etc.

This book has set out as context for that question what we human life units are and what we do. That context points toward what we might attempt to do.

The first three chapters of this book set out a basic framework for understanding "order", of which we are a manifestation. The information in the third chapter on characteristics of the ordering process can help us better navigate in this ordering process.

The fourth chapter adds more navigational assistance, by outlining how the ordering processes of the universe build hierarchies, and showing that "emergence theory" and "hierarchy theory" can in core substance be brought into equivalence with the hierarchical ordering process.

The chapters on the arrow of time, causation, and complexity are not quite as architectural, in terms of what humans do and might do, but do provide some concepts which may be useful.

The "Life" chapter puts the life process in this framework, showing that life is a product of correlational processes in "non-equilibrium thermodynamic" situations, operating off energy flows and reproducing in ways leading to evolutionary proliferation of ordered life systems. This chapter and the energy chapter depict life as channeling energy flows, exploiting potential energy arising from differentials built into the Universe, and building order in doing so.

The life chapter also shows how hierarchy building is manifest in the life process, and places humanity on life's organization chart as a manifestation of language mediated group organization among large multicellular animals. This sets the stage for what humanity appears poised to attempt at our current point in evolutionary history.

The chapter on energy sets the framework for energy potentials available to life, and human civilization, and points out that humanity faces a make or break transition point, in developing energy flows through its own social fabric as well as the legacy life system on Earth.

The globalization chapter focuses on how human energy channeling and hierarchy building have created a potential for a globally integrated human social system, which might foster human life potentials, and also help control human impacts on the legacy life system which could otherwise diminish both the legacy life system and our own human potentials.

This chapter speaks to a proposed long term orientation for humans as we address these and other challenges.

Recent public discussion may support attempting such an orientation. Scholars, and others, have begun to speak of an 'Anthropocene'. By this they mean, in general, that humans have been making such large changes in materials found on Earth, in vegetation patterns, in animal and plant survival rates, and in abundant, widespread artifacts (like cities and all that is associated with them) that scholars of the far future (if there are any) could recognize that we are noticeably changing the Earth's surface. And that easily leads to the question about where humans are going with this civilization development. What are we creating? What will amount to, for us, for our descendants, and for the Earth which those descendants will inhabit?

I have offered you some rather broad themes, or frameworks, for addressing such questions. For example, "exploiting the "non-equilibrium thermodynamic" potentials for order building on our earthly life base", does sound a bit wonkish. However, we can translate that into some understandable, if still rather broad, themes.

Chaisson has given us some concepts and tools we can readily appreciate and use. He also has shown that we fit into a cosmic evolutionary process. He has shown that life units like us channel energy to create energy density and complexity. We can use measures of energy density like his to measure our current status and to express quantitatively where our civilizational constructs may seem to be going. (We can also use derivative or related formulae to measure complexity in various settings.) We can use information-theoretic tools provided by Boltzmann, Shannon, Barabasi, Ulanowicz and others to measure degrees of organization in ecosystems, economic systems, and other human and human related systems. We can use social network analyses, which are derived from relational ways of looking at the ordering process, to characterize our social and economic systems.

We have intelligible concepts deeply grounded in empirical sciences, and a basis for deriving empirical, quantitative sciences to use to apply those concepts. This is a much better conceptual position to be in than was available to us in the foregoing centuries, including the 19th and most of the 20th centuries.

So, I propose that we are in a better position to address meaning-of-life questions than in preceding centuries.

But I also propose that we should and must approach our orientation in a much less parochial, less human-centered way than we have to date.

I propose that if we want to continue to live the high energy, high complexity lives we have recently become accustomed to, we do need to develop the 'renewable' energy systems we are now putting so much effort into, largely replacing fossil fuels systems, at global scale.

This is obvious, in our own interests. But it is also a potential extension of the basic life process of accessing potentially energy sources in a differentiated universe, and building order out of those energy flows. We can be seen as an extension of life's basic nature, and an engine of life's development.

But I wish also to suggest that we must help build the Earthlife system as a whole, not just the human part. In attempting to do so, we need to be guided by what our legacy lifesystem can sustain and what it can be developed to sustain.

A schematic basis for this proposition which I here offer is a bit abstract. It does, however, depend upon concepts introduced before.

Here are the keys to the argument. We recognize that complexity is related to energy flows and energy densities. We recognize that our civilizational construct is an high energy density, high complexity construct. And we recognize the ubiquity of the power law distribution of events in the universe, including its application to things like complexity.

With these things in mind, let's look at an interesting construction which Stephen J. Gould has offered – though he offered it with quite different ideas in mind, I think.[65]

In his book Full House, Gould uses a curve to show the distribution of complexity in a lifesystem. I reproduce the essential idea below.

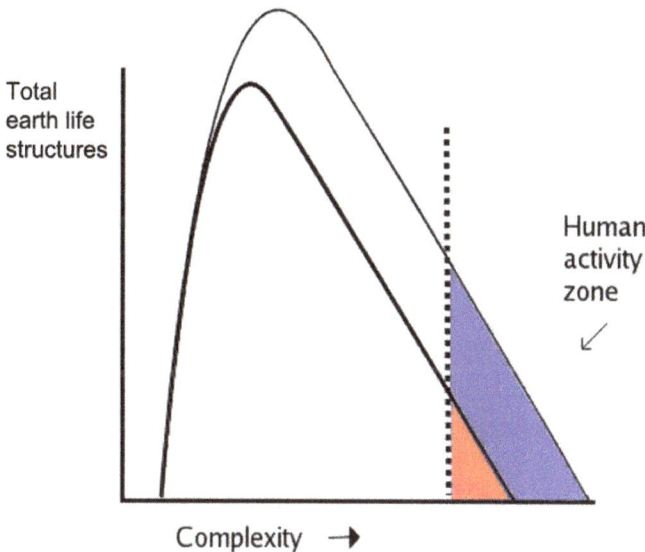

Fig. 11.1: Future Scope for Human Activity Zone.

---

65 Gould might take exception to the liberties I am taking with his conceptual apparatus. But that what happens when you publish. You just can't control what people are going to do with the stuff you put out there.

In this curve, in his original construct, Gould in effect argued that the number of organisms is scaled from bottom to top, or vertically, and increases in complexity are scaled from left to right, so the highest complexity zone is in the little tail on the right.

This curve in effect argues that in an ecosystem you have lots of relatively simple systems, or organisms, and relatively few complex systems, or organisms.

Gould used this construct to argue that life did not, over evolutionary time, alter the shape of the curve. Over time, Gould argues, you did not get a higher proportion of complex things to simple things.

However, if you hiked up the curve – got more things under it – then you could get a greater total quantity of complex life forms, in proportion to the greater total quantity of all life forms. You can see that illustrated on the graph.

Two key assertions underlie this graph. I want to make them explicit.

First, Gould is in effect assuming that there is a curve which expresses the potentials for complexity in the ecosystem, and the shape of that complexity curve is not going to change.

If a power law applies to complexity, this assumption is correct. I have argued that a power law curve must apply to complexity measures. Thus Gould and I are on the same page, whether or not we draw the curve with the same axes and shape as Gould uses in his illustration.

In addition, however, I have argued that the shape of the curve is invariant – we cannot change its shape. It will be a power law curve.

Secondly, Gould is in effect arguing that for there to be highly complex life elements there must be a substructure of less complex ones.

That is, Gould is assuming that there must be lots of microbes, and small metazoans, for there to be a relatively small number of large complex multicellular life forms. The complex stuff is embedded in and supported by the less complex stuff.

Again, I agree. There is considerable evidence to support this assertion. It is consistent with – though not necessarily obvious from -- the "mutual information" pictures of Ulanowicz which I have briefly described earlier.

If we accept these premises, Gould has given us a convenient way of representing a concept with some very important implications. [66]

Let's suppose that human life is over on the right hand side of this curve. We are a large component of the more complex part.[67]

---

[66] Here is an interesting question which the Gould notion raises. How much complexity can a life colony support? If you have a life curve on the Moon or Mars with little space under it, how much human and technological complexity could the colony, alone, support? Or a sealed-off spaceship, including the familiar science fiction (life) Seedship?

[67] This supposition may offend many. I get the clear impression that many resist this notion. They have some reason to do so, I must concede. A species which for millennia had the sun going around an earth of which we were necessary lords and masters dislikes humbling. For our own good, we need

Now we come to the point of my argument. <u>If this construct is a usable guide, for there to be more of us and our artifacts, there must be more life structure.</u>[68]

However, in doing this, I find that, conceptually, I have to expand Gould's concept to include all the life-related systems which support life. This seems compelled if we consider that human life must build systems which entrain more energy into the life complex, to support the civilization we add onto the legacy lifesystem – dams, roads, windmills, nuclear power plants, etc. Also, at the legacy lifesystem level, we can recognize that the oxygen in the atmosphere is a creation of life and supports its more active components.

In this extension of Gould's concepts, we have to hike up the curve to include more life structure underneath it, for us to have a greater scope of activity.

Either that, or many of us will have to find low-complexity embodiments and lifestyles.

Do you want to live a life about as complex as that of a cow? Not to deprecate another species, but I think I would prefer more action.

Let us go back to the concept that earthlife continues to run off a two track energy flow system -- that is, the legacy biological photosynthesis system, and in addition our artifactual energy supply system or systems. Thus, I argue, the human-built artifacts would need to be considered as components in any future ecosystem.[69] Such artifacts can be relatively simple components, as well as complex components.[70]

Conceivably the cyber-ecology component might be larger than the pre-human biological ecology, but the biological component might arguably be no larger, or be smaller than the artifactual part.

However, I, and I think many others, would suggest that we take care that we do not crowd out, or down, the legacy system. The legacy ecosystem is not only grand, majestic,

---

to cast a dubious eye upon, and test thoroughly, assumptions of exceptional status. However, let us look at the energy and complexity constructs which Chaisson offers, and recognize that our high energy lifestyle embodies a degree of complexity not evenly spread out in the ecosystem.

68 What does "life structure" consist of? I would suggest that the structure of elements which underlie and support biological units would be a part of this concept -- oxygen in the atmosphere, the calcified portions of coral reefs, highways, buildings, the inanimate parts of termite mounds, etc. To argue that the life structure must be enhanced does not necessarily imply that the total biomass be increased. But given the intimate linkages in the biosphere, there is reason to suppose that this may well be the case. And there is reason to presume that if we do not take care of a robust biosphere, all our prospects may be diminished.

69 This is easy to see if you look around you today. Have you noticed, looking down from an airliner, the large green circles on arid western lands where wheel irrigation us used? Or look at the man-made lakes, behind our dams. When our descendants look at satellite images of desert lands a millennium from now – assuming our society holds together well enough to have satellites – will they see seas of blue tinged solar energy absorbers?

70 We all know that many human artifacts appear complex to us, and the rapid development of computers presents us the possibility that the human brain will lose its (not fully established) claim to be the most complex 1200-1500 cubic centimeters in existence.

and apparently very rare in the universe, it did create us, and is and will continue to be our basic support. It would be foolish to weaken or collapse our foundations.

You, the reader, may or may not be inclined to agree with this Gould-related construct which I offer concerning to create a large ecosphere to house our complex human society. Some friends have suggested some arguments against reliance on this curve construct for the conclusions I have drawn from it. I do not at present find these arguments persuasive, but set them out in the footnote.[71]

Thus, I offer you a straightforward argument for mutualistic interactions between humanity and Earthlife as a whole. If we want to expand our scope, we have to expand the scope of Earthlife as a whole. The corollary implication is that if we degrade what might be called the ecostructure as a whole, we will have less scope for the high-energy complex life style for ourselves which we have come to love.

This is not a very easy or simple challenge. If we are to create or mediate a larger life structure, we introduce major management – or co-evolution -- issues, some of which we are dealing with today.

Using fossil fuel systems generating less than 10% of the photosynthetic energy flow on Earth, plus our appropriations from the photosynthetic system, we have produced major crowding out effects in Earth's ecosystem. We are also, global surveys suggest, pressing hard on the legacy system's ability to sustain our increasingly large and active human population.

---

71 One argument is that we can change the shape of the curve. This argument might go as follows. We have come up with fossil fuel – mining life's storage of energy. In doing so, we have not noticeably expanded the legacy ecosystem. We have thus arguably built much more complex activity than previously existed without expanding the legacy ecosystem. Thus we must have changed the shape of the complexity curve.

But if power law measures are compelled by the nature of the universe, we must fit within them, sooner or later. One can imagine that we have – at least temporarily -- expanded the life structure we tap – by adding the fossil fuel portion. Or one can argue that our complexity bulge – our increase in complex activity -- will not be sustainable. Many have argued unsustainability.

Another challenge is to ask what we make of the fact that some complex phenomena might be considered as energy minimizing aspects of dynamic activity. All reductions in variance can be seen as minimizing energy, within the reduced-variance system. In less abstract language, can we not suggest that social insects, and other groupish life activities, internalize energy flows in life structures more efficiently than less-groupish activity patterns? Ant hives process energy more efficiently – turn it into biomass more efficiently – than disorganized insects do. So, the argument would go, do human hives.

I myself have suggested this view. Others have hinted that ordering activity may minimize energy flows, or make more efficient energy flows. So, an argument might be offered that if you jack up energy flows, efficient uses of those flows will take manifestations of complexity off an hypothetical log normal curve.

But my response would be that these energy efficient developments in the complexity curve are an inherent aspect of the curve itself. Energy flows allow for complex, efficient manifestations of structured energy. Energy flows allow for energy density. However, the complexity curve does not and will not change its shape.

If we use the co-evolution frame of thought, and use it to probe means of win-win relationships with Earth's ecosystem, we would, while pressing ahead with our "alternative energy" programs, take great care to try to avoid systemic loss in our incredibly rich and unique earthlife legacy beyond our ability to remedy.

That is, if and as we re-route rivers, pave the desert with sunlight absorption technologies, suck velocity out of the wind, split atoms, elaborate and enrich our urban human hives, we would need to be asking ourselves whether we are making a net addition to the scope of earthlife, or impoverishing it by the waste of evolutionary treasure which we do not know how to replace.

How is the perspective which I suggest different from those used by governments today, and even by environmental protection groups whose aims are similar to those I suggest?

To put the issue in contrasting terms, many governments and environmental groups speak in terms of preserving environmental services to humankind. This is right, but also wrong. I suggest we think and speak in terms of humanity augmenting earthlife.

We want more and larger human hives. To have them, we will need to have mines, and transport systems, desalination plants, and dams -- in the context of forests, and grasslands, and and active biological carbon sinks and storage.

So now we come to the "ought to" questions. If our long term best interests, on a global basis, lie in helping create a more ample earthlife, as an extension of the Universe's ordering process, what ought we humans to do in our various levels of organization, our various engagements?

This can be seen as an "ethical" issue. "Ought to" issues often are thus seen.

Let us put the much-worked concept of ethics in the correlational framework set out in this manuscript. We see the ordering process as creating organization out of constraints. We can consider "ethics" as a type of constraint on the actions of an element -- an animal element -- in a field of relationships. If we generalize this concept a bit, to see a protocol as an ordered – and thus constrained – set of actions, we can see ethics as protocols for actors to follow so as to achieve effective group organization.

Those protocols require us semi-autonomous agents to forego degrees of freedom. In our accustomed human perspectives and terminology, when we commit to integration in a collective, so as to provide benefits to the collective, and derivatively its constituents, we also give up options, or actions, some of which might provide some benefits for us.

An enormous amount of work has been done on whether, why and how "ethics", or altruism, has evolved in humans and other organisms. I will briefly touch upon such work in this chapter, But I offer a deeper organizing concept.

The efficient capture and efficient use of energy flows provides, at bottom, the benefits to group activity. This is the source of any continuing, sustained payoffs which might be postulated in models entailing group benefit. The structure and function of living organisms, and social systems of organisms, are shaped by the energy flows on which they subsist. Their internal coding systems, their 'genes' and associated systems, have been shaped in life's long evolutionary process, to build in constraints and action potentials which effectuate acquisition of energy flows -- and in the case of 'social' creatures, to effectuate group acquisition and deployment of energy flows.

Thus it must be with any human social system, global or otherwise. The coding systems for human social systems have been and will be shaped by these imperatives. Thus, "ethics", or protocols, will arise which serve group function requirements, at each and all levels of social life, and social human life, organization.

Often in human terms we see constraints as "rules" and "regulations", or expectations (or any of many equivalent terms we use). Ethics and laws, or rules, overlap considerably. One can consider laws as formally codified ethics, backed up by the force of a formal organization, such as a corporation or the State. Not all "ethics" are so codified, or backed up the coercive force of a monitoring organization.

Many protocols, or rules and regulations, appear to us as likely to result in personal well being, and we accord with them without protest or a sense of internal objection or difficulty. But sometimes the sense of loss, or the reality of loss, is palpable and significant to us, or another actor.

We tend to consider an issue an "ethical" matter when it appears to an actor that the constraint will lead to some loss of significant potential gain which might otherwise be available. For a few examples out of an infinity, we give up the chance to gain from deception in or welshing on a bargain, to dodge a governmental tax in a way which might be called legal but is perceived to be abusing a collective social bargain as to citizen tax sharing, to speed our walk on the city's streets by leaving our dog's droppings instead of picking them up, or, at an extreme, to give up one's life for another, or for a collective.

Among those who study evolution, the question has often put in terms of "altruism". Scholars have addressed the question how such self sacrificing behavior could have evolved among creatures whose evolutionary standing rests on reproduction, when the "self sacrifice" amounts to giving up individual reproductive advantage

This investigation might seem a little odd, or at least constricted, when put squarely in the context of a demonstrable track record of life creating hierarchies of groups, from single celled organisms, and even precursors of them, up to social multicellular organisms.[72] But it addresses a "free rider" sort of problem visible to humans in our own socialization system, it treats with important mechanics in evolution by use of biological code strings, or genetics, and it carries in it a powerful explanation for kinship preferences which are obviously pervasive in human socialization.

In my opinion at the time of this writing, Martin Nowak, at the program for Evolutionary Dynamics, at Harvard, has produced the best general set of theses, among the traditional analysts of ethics and altruism in life organization.

---

72 To pose this question in a bit more balanced fashion, one recalls that a lot of solitaries are left out of groups, at every stage.

Nowak is associated with the view that the emergence of genomes, cells, multicellular organisms, and human society are all based on cooperation. Cooperation enables evolutionary dynamics to be constructive, in his constructs.

What we humans call cooperation is obviously a form of correlation, and, understood in this framework, Nowak's observation is accurate.

Nowak (2006) has proposed five categories for reward for "altruism" considered as group-favoring cooperation -- kin preference; direct reciprocity; "indirect reciprocity" which is mediated or facilitated by elements within a group having a reputation for being a good reciprocator; "network reciprocity", where cooperators form network clusters where they help each other and thus give each other a higher probability of reward; and "group selection", which deals with the realization that a group with better cooperation systems, or coordination systems, can out-compete a groups with less coordinated processes. Nowak's group has been able to work out some formulae which could give a social engineer cost/reward targets. Nowak (2011) has also published a book entitled "Super Cooperators" in which he presents his history and findings for the general public.

Current human social systems embody easily perceived systems designed to enhance the probabilities of benefit from agent or element actions which confer benefits on "group" or "cooperating" elements. One can observe that many human institutions seem to be arranged to increase the likelihood of payoffs to individual actors from participation in group-supporting activities -- that is, in a sense, increasing the likelihood of mutual benefit reciprocation, direct or indirect, from "ethical" activity.

To repeat, the efficient capture and use of energy flows provides, at bottom, the benefits to group activity. This is the source of any sustained payoffs which might be postulated in models entailing group benefit. The structure and function of living organisms, and social systems of organisms, are shaped by the energy flows on which they subsist. So it must be with any human social system, global or otherwise. The coding systems for human social systems will, accordingly, be shaped by these imperatives.

This means we must try to fit our ambitions into a range and types of energy flows for which we can devise advantageous incorporation and use, in a "goldilocks zone" -- neither too much nor too little, at any given time, sustained and sustainable, and if possible growing.[73]

As we have interaction between elements in any given system, and volitional human activities which affect such interactions, we have seen, ubiquitously, what we

---

73 Obviously much of the concern over global warming relates to whether we will overheat earthpark. Past greenhouse gases, if we get past them successfully, we would need to avoid focusing so much energy into our activities by non-fossil means as to overheat the earth in new and ingenious ways. And to the extent we can affect the temperature of the Earth's surface, we would face the political feat of holding together some coherent means of governing our doing so successfully over millennia.

call networks. In this framework, we have come to consider the durability of networks, at various scales. And so considerable attention is now given to questions of network integrity, in internet situations, say, and in societies. What happens if you take out a node, or a type of node, or a set of nodes, etc.? In the recent financial crisis, we got a real-world demonstration of what can happen when linkages in the global economy fail, and networks of interdependent trust and confidence contract. So also do money flows, economic activity, and the general sense of well being of the population as a whole.

So as we consider human and non-human complex systems, we will need to be coding for network sturdiness, or resilience.

Coding for network integrity is, however, closely related to the traditional categories of ethics and altruism. If we generalize the extant modeling on altruism, we can say that any code element in a system (we can call a gene or a gene carrier as a code element, expressing a function within a network), requiring energy flows for sustenance, will experience a linkage, or lack of linkage, between its expression and the maintenance of its place in a network. Whether the "reciprocity" is "direct", or "indirect", or a part of network dynamics, there must be correlations, or regularities, in the dynamics in which it participates which sustain it. Further, in tiered life hierarchies, coding at one level reflects and may affect regularities or lack of regularities at higher levels. Genes code for interactions between organisms, social codes at the state level have to reflect conditions at an inter-state level, etc.

The foregoing discussion suggests parallels between genes and laws – or codified customs. We can see genes as code systems, or code elements, and we can see laws, procedures, customs as human-assembled code systems. The parallels are numerous and close.

Just as genes may be seen as evoked and tested by the potentials for and constraints of life operation, so may we consider our human codes. Just as gene systems evolve to reflect mutual information with the environment, so must our human culture/ organization codes. Just as genes must, to be maintained, be involved in interaction patterns which feed energy into the organism in which the genes are housed, so must human codes, at various scale, code for interaction patterns which maintain the structures in which they are housed. Just as genes code for action patterns which have effects in hierarchical levels above the cell in which they are housed, so will human codes affect the action patterns of human organizational units which are subsumed in inclusive hierarchies. Human codes will specify the interaction patterns between hierarchical levels and between an organization at each level and the elements external to the organization with which the organization interacts. Just as genes code for contingencies – if ambient temperature for a dog is warm, then the hair shedding pattern is activated, if cold the hair growing pattern is initiated – so do and will human organization codes, such as legal codes.

Human societies have a wide range of competencies. We are at an immature stage of international organization. We will be very fortunate if we can get to a global ecumene having predictable, functional, productive norms, or, to put things in lawyer-speak, a de facto productive international rule of law.

So we have ongoing social engineering efforts, with some clear issues before us. Can we identify one or more over-arcing themes we need to address?

I have suggested several themes in the chapter on globalization. But a more basic, underlying theme is first and foremost, though also perhaps the most difficult.

At this point, we need a global orientation to our global potentials. The global orientation I propose is this: we need to stop seeing this good earth as a set of services for humankind, and instead see humankind as a trustee of the potentials of life on this Earth.

What is in this vision for us? If we maintain scope for realization of life potentials on Earth, we maintain scope for our own potentials, as argued before in this chapter.

I do not hear any major leaders among us calling for this vision. They are constrained by our inherent, dogged, self centered parochialism. We have learned that the universe does not revolve around the Earth. But we are still determined that the Earth revolve around us.

But we cannot know this Earth's possibilities, and how we may best fit into those possibilities, if we do not root out this last, tenacious parochialism, this infantile view of Earthlife.

To paraphrase Kabril Gibran's claim that our children are not our own, our future is not our own, nor ours to own. The future is a set of potentials set before us by the universe. With our collective language and imaging capabilities, we have been given the means of understanding much of how the universe works, including, if we will, how we might act to realize those potentials.

Thus we must determine what this earthlife creation of large, social, communicating mammals, ourselves, can do to enhance the energy budget of earthlife. We must use the treasure trove of stored energy of life past to expand the current and ongoing energy budget of life on this Earth. As we do so, we must learn how to harmonize new energy flows with the legacy photosynthetic energy flows, rather than merely parasitize the legacy photosynthesis system.

In the terms of "emergence", we must find ways to discipline the emergent effects of human socialization on Earth, and its life. As so many among us have urged, our ethic, our constraints, must in part be the preservation of the foundations of our life setting, and thus the potentials of our companions in life's odyssey.

We must devise the protocols which allow us to make of our many local and regional organizations, our kinship groups, our many hard-won but limiting parochialisms, an interconnected, well conceived, productive, stable, uncorrupted and unparasitized, sustainable and sustained global ecumene.

Fundamentally, we must learn techniques for organizing the resources for life on earth, so that life here may more abundantly proliferate. We have already been reshaping the earth. We must devise a vision of abundance in life realization, rather than mean, stunted subservience to us referring only to our self-perceived current needs. And we must learn prudence and patience in the service of that vision.

Why "must" we attempt these extraordinarily difficult and demanding things, with no guarantee of success?

I submit that the most grievous costs are opportunity costs -- the failure to realise what this Universe makes potentially available to us. If we do not address the world in these terms, we will have no way of knowing what possibilities exist and can be made to exist for earthlife, and for us in it. Only if we conceive ourselves as trustees of life potentials and commit ourselves to this vision can we know what this magnificent Earthlife legacy can become, and what we can become, within it, serving its potentials, in fruitful accord with our universe and our Earth home.

Appendix A: Note on Ontology and Epistemology

This treatment of the nature of order, and thus of the sensible universe, which assumes relational quantum mechanics and the basics of the "Standard Model" and general relativity -- though it raises questions about the Second Law as traditionally understood -- is rather spare in its ontological and epistemological foundations. This may be seen as both a vice and a virtue.

First, the fundamental concepts -- correlation, differentiation, relation, emergence and combinatorial elaboration -- are self referential and almost circular. Each is explained by the other.

All, however, are set on a quantum mechanics foundation, in which quantum systems "measure" each other. These "measurements" are relational, in that they express the relationship between the quantum systems. As Rovelli pointed out, such encounter, or relational, acts, are ubiquitous throughout the system which creates a universe, as we understand it. I have added to that concept that they comprise, as they create, the Universe as we know it.

These fundamental concepts are made operational with reference to invariances in the operation of the Universe which we infer from observation -- the Planck length, the "speed of light", the conservation of energy, symmetries to the extent observed, etc.

There may be some question as to whether any given invariance which we perceive and reflect in verbal and mathematical formulae is "fundamental", or is instead a construct of more fundamental things. There may be aspects of the initial expression of the universe, or its eventuation, which would not manifest such invariances, since we know so little about such things. But they are observed in the rigorous procedures of "science" in the current function of the universe. The history of the universe as we can trace it suggests they have been in effect for as long as we can trace.

The basis, I suggest, for this self referential and tautological foundation is that our explanation of our universe is an act of an element of the universe engaging in explaining itself by itself. Our cogitation is a recursive act. We use the fabric of which we are made, the modes of our existence, to explain ourselves and our universe, because we have no other tools with which to function.

Thus, this explanation of order in this universe does not rest upon any apprehension of explaining what became before this Universe, or what might have what we would call "existence" "outside" this universe. (Again, one may refer to the interesting discussion by Tegmark, earlier cited, and the thought experience about a 'gas' universe.) The implication of this is that this universe may be considered by us, its creatures, within its own framework, as singular, and historical. If we try to construct a "framework" for this universe outside, prior to, or in some way other or more inclusive than this universe would seem to depend upon the ability to construct such a framework working from the internal mechanics of this one. How would we validate that?

This self referential foundation has its obvious limitations, which have frustrated inquiring minds from time immemorial. But, paradoxically, it may also attest to the validity of the exercise, i.e the fundamental nature of the exercise, if the concepts do serve to track what is going on insofar as we can observe it.

In Lee Smolin's language, (2014), pp 108-113, this theory of order is proposed as an "effective" theory, without meeting all the conditions, or goals, he envisages for comprehensive theories of the universe.

I, among a great many others, would like to peer beyond the curtain of the Initial Cosmological Event, and the "Initial Conditions". Lee Smolin, and others, offer many stimulating suggestions. But I read such exercises with reservations and here make no effort to prove what may lie beyond these boundaries.

Notwithstanding this, "string theories", to the extent I dimly apprehend them, do offer an interesting perspective. That self consistent symbolic constructions can be made which have some correspondence to what we experience, and also be made to hypothetically correspond to what we do not experience in this universe, challenges the ontological and epistemic closure which I have here accepted as a basis for navigating the universe we do currently experience.

Also, Smolin (2014), at pp 406 et seq., indicates that DeWitt-Wheeler equations can be taken to support a universe "bounce", process, with a preceding universe closing down, so to speak, and our universe appearing on a bounce.

Recognizing the great danger of making meaningless statements out of ignorance, I do conjecture whether the admissible scope of such constructions as might be applied to "universes" might be usefully be defined as those string theories which allow for the forms of correlation which Carlo Rovelli sets out in his seminal work on relational quantum mechanics (as well as the Pauli exclusion effect, of course).

The explanation of what our "universe" is, here offered, would limit the use of the term "universe" to such realizations of string theoretic constructs.

Thus, I do not here attempt any judgment on "time" or process of the sort we understand before or after our Universe. Qualified persons are busy on these unanswered questions. This essay is addressed to the nature of the observable, tangible order in the universe which we now experience, at the current stage of its realization.

Appendix B: A Selection of Human Social Conventions Which Favor Cooperative, and Thus Correlated, or Coordinated, Behavior

Contract law has evolved to mediate and to enforce reciprocal bargains, in one on one, one on many, and many to many situations.

The bulk of the consumer protection laws in the United States are designed to require accurate information as to goods and services offered in bargain. The criminal law of frauds deal with extreme failures of information accuracy.

Regulations as to labeling, grading, efficacy and safety of consumables lessen information search costs, and enhance degrees of combinatorial fit – tending to improve the probability and degree of benefit in social transactions.

Product standards – public, private, and quasi-public – also have evolved in such a way as to increase the efficiency of combinatorial arrangements, and in doing so to increase the reliability (read probability) of mutual benefit transactions.

"Ratings" organizations – e.g credit rating organizations, and financial instruments and corporation rating organizations – convey information on the reliability of promisors, or entities cooperating in bargains of various sorts.

Human societies also embody other arrangements which reduce risk and burden of altruism, so to speak. These systems include prizes for new innovations, patents for innovations, social recognition for social service, pensions for wounded soldiers, subsidies for services thought to benefit the collective, etc. "Public servants" often do well financially. "Social service" in a modern society often brings recognition, esteem, and a pretty good living.

Currently, at the international level, World Trade Organization standards for accession, or entry, embody a number of requirements which tend to limit predation in, promote opportunities for, and promote regularity and reliability in economic transactions. Some of the requirements for entry into the European Economic Union can be said to have similar functions.

Appendix C: Quantitative Wikis For Energy Supply and Distribution Modeling

As noted in the chapter on energy, worldwide, a very substantial number of people and organizations are attacking the energy supply future from a wide variety of positions and perspectives.

In "Western" polities, there has been a strong bias for letting "the market" work out such a discovery process, avoiding "central planning" and "industrial policy". This preference has roots in both the nature of the discovery process and in experience with both markets and central planning.

The number of potential combinations of factors of production and the complexity of the interplay of all the factors I have listed above makes difficult prediction of optimal arrangements. Systems of laws, commercial exchange possibilities, and flexible capital markets have, in the "West", facilitated efficient market discovery of innovative and productive production processes, in many fields of economic endeavor, over many decades now. Markets are, among other things, combinatorial machines.

But in practice government agencies have long been engaged in "basic research" and in trying to supplement – and sometimes facilitate – market developments. We have a long tradition of basic and process research in agriculture, energy, computing, and other sectors.

In recent years, a sense of urgency in getting additional and diversified energy supplies on line has been pushing government bodies in many countries and at several levels toward trying to stimulate and steer "alternative fuels" production. It appears at the time of this writing we are going to see more of that in the United States in the years ahead.

One possible way to assist this discovery process arises from the advance of computational capacities and lowering of communications costs. We may be able in effect to "wikify" the policy analysis, and risk analysis, tasks – the task of identifying potential paths to energy development and identifying potentially useful points of public intervention or action, as well as private investment.

As noted above, we have a host of contenders for power supply – solar, geothermal, biofuels, wind, nuclear, wave, fuel cell, etc. Each area has sub-areas (e.g. in photovoltaics contending technologies include silicon cells, a variety of thin films including dye based films, concentrators, tower-mirror arrays, etc.) Each area has to fit into the complex web of existing energy uses and energy delivery systems. In today's environment, each area and the sub-areas are candidates for public investment, at several levels.

Though these combinatorial possibilities are likely to be sorted out by decentralized search efforts by a host of "private" and "public" actors, modeling aids might be considered.

One has to do with the sort of input/output analyses pioneered some decades ago by Vassily Leontief, and developed in various ways since. For example, in 2016 a company called Otherlab has created a very detailed chart of energy allocations to use sectors in the United States.[74]

This sort of analysis leads to thinking in terms systems of nodes and flows, resembling the ecosystem charts of Bob Ulanowicz. In the age of the internet one might consider putting on the web the formal structure of matrices which would allow tracing out the effects of alterations in demand conditions, distribution arrangements, technological possibilities, changes in relative costs, etc. Such as structure could, in effect, with proper computational inclusions, constitute a substitutability matrix, allowing tracery of alterations in costs and output throughout the interconnected fabric.

The matrix could identify areas of significance to development and deployment in each sector – e. g. drilling costs for geothermal energy, cellulosic degradation potentials and algae taming in biofuels, electric transmission deployment for wind and nuclear, etc.

One would allow for response and tweaking, allow for research and other groups to plug in results/proposals, and allow for user modeling of scenarios. One might have senior editing supervision of contributions, and allow for general public preference ranking.

Such a facility would in effect simulate on the public stage, available for wide viewing and input, the issues in the policy debates in government venues, and also simulate to some extent what the market does in less-stressed times when there is not so much pressure for collective public intervention and guidance. Such a facility would tend to bring into play a wide range of sources of information, as does Wikipedia in its operation.

A number of types of entities might undertake to entrepreneur an initiative or initiatives of this sort – universities, research institutes, our National Academy of Sciences in the United States, government bodies (e.g. departments of energy), trade associations, even "private sector" firms or consortiums of firms.

Such a system might be structured along the lines of a quantitative wikipedia-like system, allowing modelling entities to post mathematical models of technical and economic energy generation and use systems online, using a toolkit of posted and postable softwares, such that model scenarios could be explored subject to editorial review.

If one or more such systems could be created, this might significantly broaden the scope and participation base of EROEI modelling, among other other forms of modelling energy supply, distribution, and use processes.

---

74 The company is called "Otherlab", and based in San Francisco. An internet illustration of its work product is available at http://www.fastcoexist.com/3062630/visualizing/this-very-very-detailed-chart-shows-how-all-the-energy-in-the-us-is-used.

References

Annual wind report confirms tech advancements, improved performance, and low energy prices. (2016, August 17). Phys.org. Retrieved from http://phys.org/news/2016-08-annual-tech-advancements-energy-prices.html

Arrow of time. (2016). In Wikipedia. Retrieved from http://en.wikipedia.org/wiki/Arrow_of_time.

Banathy, B. (1996, December, to 1997, December). *A taste of systemics*. Paper presented at the First International Electronic Seminar on Wholeness. Retrieved from http://www.newciv.org/ISSS_Primer/asem04bb.html

Barabasi, A.-I., & Frangos, J. (2002). *Linked: The new science of networks*. Jackson, TN: Perseus Books Group.

Bent, R., Orr, L., & Baker, R. (Eds.). (2001). *Energy, science, policy and the pursuit of sustainability*. Washington, DC: Island Press.

Bhandari, K. P., Collier, J. M., Ellingson, R. J., & Apul, D. S. (2015). Energy payback time (EPBT) and energy return on energy invested (EROI) of solar photovoltaic systems: A systematic review and meta-analysis. *Renewable and Sustainable Energy Reviews, 47*, 133–141. Retrieved from http://astro1.panet.utoledo.edu/~relling2/PDF/pubs/life_cycle_assesment_ellingson_apul_(2015)_ren_and_sustain._energy_revs.pdf

Bloomberg, L.P. (2016). *New energy outlook 2016: Powering a changing world—Executive summary*. Author. Retrieved from http://www.bloomberg.com/company/new-energy-outlook/

Brooks, D., & Wylie, E. O. (2006). *Evolution as entropy*. Chicago, IL: University of Chicago Press.

Buchanan, M. (2000). *Ubiquity: Why catastrophes happen*. New York, NY: Three Rivers Press.

Buss, L. (1987). *The evolution of individuality*. Princeton, NJ: Princeton University Press.

Chaisson, E. J. (2001). *Cosmic evolution: The rise of complexity in nature*. Cambridge, MA: Harvard University Press.

Collier, J. (1986). Entropy in evolution. *Biology and Philosophy, 1*, 5–24.

Corning, P. A. (2002). A venerable concept in search of a theory. *Complexity, 7*(6), 18–30. doi: 10.1002/cplx.10043

Dawkins, R. (2016). *The selfish gene: 40th anniversary edition* (4th ed.). New York, NY: Oxford University Press.

Deudney, D., & Ikenberry, G. J. (2009, January/February). The myth of the autocratic revival: Why liberal democracy will prevail. *Foreign Affairs, 2009*(January/February), 77–93.

Dostrovsky, I. (1988). *Energy and the missing resource*. Cambridge, UK: Cambridge University Press.

Drake equation. (2016). In Wikipedia. Retrieved from https://en.wikipedia.org/wiki/Drake_equation

Ellis, G. F. R. (2006). On the nature of emergent reality. In P. Clayton & R. Davies (Eds.), *The re-emergence of emergence*. New York, NY: Oxford University Press.

Fossil Museum, retrieved from http://www.fossilmuseum.net/Tree_of_Life/Stromatolites.htm.

Food And Agriculture Organization of the United Nations, retrieved from http://www.fao.org/docrep/w7241e/w7241e06.htm#2.1%20photosynthetic%20capture%20of%20solar%20energy

Garrett, T. J. (2011). *How persistent is civilization growth?* Salt Lake City, UT: University of Utah. Retrieved from http://arxiv.org/pdf/1101.5635v1.pdf

Garrett, T. J. (2012). No way out? The double-bind in seeking global prosperity alongside mitigated climate change. *Earth System Dynamics, 3*, 1–17. Retrieved from http://www.earth-syst-dynam.net/3/1/2012/esd-3-1-2012.pdf

Garrett, T. J. (2015). Long-run evolution of the global economy: Part 2. Hindcasts of innovation and growth. *Earth System Dynamics, 6*, 673–688. Retrieved from http://www.earth-syst-dynam.net/6/673/2015/esd-6-673-2015.pdf

Globalnet summary of global net primary productivity statistics, retrieved from http://www.users.globalnet.co.uk/~mfogg/icons/calc3.html

Goodenough, U., & Deacon, T. (2008). The sacred emergence of nature. In *The Oxford handbook of science* (Chapter 50). New York, NY: Oxford University Press.

Gordon, Robert J. (2016), *The rise and fall of american growth*, Princeton University Press.

Gould, S. (1996). *Full house*. New York, NY: Random House. (Republished by Harvard University Press, 2011).

Hall, C. A. S., Lambert, J., & Balogh, S. B. (2014). EROI of different fuels and the implications for society. *Energy Policy, 64*, 141–152. Retrieved from http://www.sciencedirect.com/science/article/pii/S0301421513003856

Hulswit, M. (n.d.). *A short history of causation*. Toronto, Canada: University of Toronto. Retrieved from http://www.library.utoronto.ca/see/SEED/Vol4-3/Hulswit.htm#_edn10

Intergovernmental Panel on Climate Change (IPCC). (2007). *Climate change 2007: The physical science basis*. Contribution of Working Group I to the Fourth Assessment Report of the Intergovernmental Panel on Climate Change [Eds., S. Solomon, D. Qin, M. Manning, Z. Chen, M. Marquis, K. B. Averyt, M. Tignor & H. L. Miller]. Cambridge, UK: Cambridge University Press.

Intergovernmental Panel on Climate Change (IPCC). (2011–2012). *Renewable energy sources and climate change mitigation: Summary for policymakers and technical summary* (Special IPCC Report. Geneva, Switzerland: IPCC. Retrieved from https://www.ipcc.ch/pdf/special-reports/srren/SRREN_FD_SPM_final.pdf

Jantzen, D. (2000, November/December). *Wildlands as garden. National Parks, 2000* (November/December).

Joos, E. (2006). The emergence of classicality from quantum theory. In P. Clayton & R. Davies (Eds.), *The re-emergence of emergence*. New York, NY: Oxford University Press.

Kauffman, S. (2000). *Investigations*. New York, NY: Oxford University Press.

Layzer, David (1990) *Cosmogenesis*, Oxford University Press.

Lenton, T., & Watson, A. (2011). *Revolutions that made the earth*. New York, NY: Oxford University Press.

Likens, G. E., & Whittaker, R. H. (1975). The biosphere and man. In H. Lieth, & R. H. Whittaker (Eds.), *Primary productivity of the biosphere* (pp. 310–311). New York, NY: Springer Nature.

Margulis, L. (1970). *Origin of eukaryotic cells*, Yale University Press.

Mermin, N. D. (1996, September). *The Ithaca interpretation of quantum mechanics*. Notes for a lecture given at the Golden Jubilee Workshop on Foundations of Quantum Theory, Tata Institute, Bombay, India. Retrieved from https://arxiv.org/pdf/quant-ph/9609013v1.pdf

Mill, J. S. (1843). *A system of logic*. New York, NY: John Parker.

Morowitz, H., & Smith, D. E. (2006). *Energy flow and the organization of life* (SFI Working Paper 2006-08-029). Santa Fe, NM: Santa Fe Institute. Retrieved from http://www.santafe.edu/media/workingpapers/06-08-029.pdf

Naam, R. (2015). *What's the EROI of solar?* Ramez Naam. Retrieved from http://rameznaam.com/2015/06/04/whats-the-eroi-of-solar/

National Aeronautics and Space Agency, net primary productivity statistics, retrieved at https://daac.ornl.gov/NPP/npp_home.shtml.

Nowak, M. A. (2006). Five rules for the evolution of cooperation. *Science, 314*(5805), 1560–1563.

Nowak, M. A. (2011). *Super cooperators*. Washington, DC: Free Press.

Odum, H. T., & Odum, E. C. (1983). *Energy analysis overview of nations* (International Institute for Applied Systems Analysis Working Paper WP-83-82). Laxenburg, Austria: International Institute for Applied Systems Analysis.

Pearce, J. (2016). The international productivity problem, the energy transition, and globalization, going forward. *International Affairs Forum*. Retrieved from http://www.ia-forum.org/Files/KAVOQJ.pdf

Relational quantum mechanics. (2008). In Stanford Encyclopedia of Philosophy. Retrieved from http://plato.stanford.edu/entries/qm-relational/

Renewable Energy Policy Network for the 21st Century (REN21). (2016). *Renewables 2016 global status report.* Paris, France: REN21. Retrieved from http://www.ren21.net/wp-content/uploads/2016/06/GSR_2016_Full_Report_REN21.pdf

Rovelli, C. (1996). *Relational quantum mechanics.* Los Alamos, NM: Los Alamos National Laboratory. The article was first written in the 1996. Retrieved from http://xxx.lanl.gov/PS_cache/quant-ph/pdf/9609/9609002v2.pdf

Ryan, J. (2016, February 4). A renewables revolution is toppling the dominance of fossil fuels in U.S. power. *Bloomberg New Energy Finance.* Retrieved from http://www.bloomberg.com/news/articles/2016-02-04/renewables-top-fossil-fuels-as-biggest-source-of-new-u-s-power

Salthe, S. N. (1985). *Evolving hierarchical systems.* New York, NY: Columbia University Press.

Schneider, E. D., & Sagan, D. (2005). *Into the cool.* Chicago, IL: University of Chicago Press.

Smil, V. (2005). *Energy at the crossroads.* Cambridge, MA: MIT Press.

Smolin, L. (1999). *The life of the cosmos.* Chicago, IL: University of Chicago Press.

Smolin, L. (2013). *Time reborn.* Boston, MA: Houghton Mifflin.

Tegmark, Max, 1997, On the dimensionality of spacetime, *Class. Quantum Grav.* 14 (1997) L69–L75, IOP publishing, retrieved from http://space.mit.edu/home/tegmark/dimensions.pdf.

Thomas, A., & Weber, J. (n.d.). Mesophases. In *Liquid crystals* (part 2). Munich, Germany: Max Planck Institute of Colloids and Interfaces. Retrieved from http://www.mpikg.mpg.de/886863/Liquid_Crystals.pdf

Tsao, J., Lewis, N., & Crabtree, G. (Eds.). (2006). *Solar FAQs.* Washington, DC: Sandia Corporation (for U.S. Department of Energy). Retrieved from http://www.sandia.gov/~jytsao/Solar%20FAQs.pdf

Ulanowicz, R. E. (1997). *Ecology: The ascendant perspective.* New York, NY: Columbia University Press.

United Nations Environmental Program report GEO 5, retrieved from http://www.unep.org/russian/geo/geo5.asp

U.S. Energy Information Administration (EIA). (2016). Electricity. In *International energy outlook 2016* [Report Number: DOE/EIA-0484(2016)] (Chapter 5). Washington, DC: EIA. Retrieved from http://www.eia.gov/forecasts/ieo/electricity.cfm

Vitousek, P. M., Ehrlich, P. R., Ehrlich, A. H., & Matson, P. A. (1986). Human appropriation of the products of photosynthesis. *Bioscience, 36*(6), 363–373.

Wiser, Ryan; Jenni, Karen; Seel, Joachim; Baker, Erin; Hand, Maureen; Lantz, Eric; Smith, Aaron, Expert elicitation survey on future wind energy costs, *Nature Energy* 1, Article number: 16135 (2016) doi:10.1038/nenergy.2016.135. Retrieved from http://phys.org/news/2016-09-tools-future-power.html

Index

www.ingramcontent.com/pod-product-compliance
Lightning Source LLC
Chambersburg PA
CBHW040139200326

41458CB00025B/6320